Direct Methane to Methanol

Direct Methane to Methanol

Editor

Meenakshi Awasthi

Direct Methane to Methanol

Edited by **Meenakshi Awasthi**

Printed in 2017

ISBN: 978-1-68117-353-5

Library of Congress Control Number: 2015939266

© 2016 by
SCITUS Academics LLC,
616, Corporate Way, Suite 2, 4766,
Valley Cottage, NY 10989

www.scitusacademics.com

This book contains information obtained from highly regarded resources. Copyright for individual articles remains with the authors as indicated. All chapters are distributed under the terms of the Creative Commons Attribution License, which permits unrestricted use, distribution, and reproduction in any medium, provided the original author and source are credited.

Notice

Reasonable efforts have been made to publish reliable data and views articulated in the chapters are those of the individual contributors, and not necessarily those of the editors or publishers. Editors or publishers are not responsible for the accuracy of the information in the published chapters or consequences of their use. The publisher believes no responsibility for any damage or grievance to the persons or property arising out of the use of any materials, instructions, methods or thoughts in the book. The editors and the publisher have attempted to trace the copyright holders of all material reproduced in this publication and apologize to copyright holders if permission has not been obtained. If any copyright holder has not been acknowledged, please write to us so we may rectify.

Contents

Preface vii

Chapter 1 Catalytic and Noncatalytic Conversion of Methane to Olefins and Synthesis Gas in an AC Parallel Plate Discharge Reactor 1
Mohammad Ali Khodagholi and Mohammad Irani

Chapter 2 N-Hexadecane Fuel for a Phosphoric Acid Direct Hydrocarbon Fuel Cell 21
Yuanchen Zhu, Travis Robinson, Amani Al-Othman, André Y. Tremblay, and Marten Ternan

Chapter 3 Syngas Generation from Methane Using a Chemical-Looping Concept: A Review of Oxygen Carriers 45
Kongzhai Li, Hua Wang, and Yonggang We

Chapter 4 Synthesis of ZSM-22 in Static and Dynamic System Using Seeds 69
Lenivaldo V. de Sousa Júnior, Antonio O. S. Silva, Bruno J. B. Silva, and Soraya L. Alencar

Chapter 5 Nickel Alloy Catalysts for the Anode of a High Temperature PEM Direct Propane Fuel Cell 85
Shadi Vafaeyan Alain St-Amant, and Marten Ternan

Chapter 6 Role of Reaction and Factors of Carbon Nanotubes Growth in Chemical Vapour Decomposition Process Using Methane—A Highlight 107
Sivakumar VM, Abdul Rahman Mohamed, Ahmad Zuhairi Abdullah, and Siang-Piao Chai

Chapter 7 Hydrogen Production Technologies: Current State and Future Developments 139
Christos M. Kalamaras and Angelos M. Efstathiou

Chapter 8	**Chemical Pretreatment Methods for the Production of Cellulosic Ethanol: Technologies and Innovations** 165	
	Edem Cudjoe Bensah and Moses Mensah	
Chapter 9	**Evaluation of Methane Yield on Mesophilic-Dry Anaerobic Digestion of Piggery Manure Mixed with Chaff for Agricultural Area** 223	
	Dong-Heui Kwak, Mi-Sug Kim, Jae-Seung Kim, Young-Youl Oh, Soon-Ok Noh Byung-Ok So, Su-Young Jung, Su-Jin Jung, and Soo-Wan Chae	
Chapter 10	**Simulation of CO_2 and H_2S Removal Using Methanol in Hollow Fiber Membrane Gas Absorber (HFMGA)** 247	
	Majid Mahdavian, Hossein Atashi, Morteza Zivdar, and Mahmood Mousavi	
	Citations 275	
	Index 279	

Preface

The oxidative coupling of methane (OCM) is a type of chemical reaction discovered in the 1980s for the direct conversion of natural gas, primarily consisting of methane, into value-added chemicals. Direct conversion of methane into other useful products is one of the most challenging subjects to be studied inheterogeneous catalysis.[1] Methane activation is difficult because of its thermodynamic stability with a noble gas-like electronic configuration. The tetrahedral arrangement of strong C–H bonds (435 kJ/mol) offer no functional group, magnetic moments or polar distributions to undergo chemical attack. This makes methane less reactive than nearly all of its conversion products, limiting efficient utilization of natural gas, the world's most abundant petrochemical resource.

Editor

114648
26 St
Scitus

Catalytic and Noncatalytic Conversion of Methane to Olefins and Synthesis Gas in an AC Parallel Plate Discharge Reactor

Mohammad Ali Khodagholi and Mohammad Irani

Gas Division, Research Institute of Petroleum Industry, West Boulevard Azadi Sport Complex, 14665-1998 Tehran, Iran

ABSTRACT

Direct conversion of methane to ethylene, acetylene, and synthesis gas at ambient pressure and temperature in a parallel plate discharge reactor was investigated. The experiments were carried out using a quartz reactor of outer diameter of 9 millimeter and a driving force of ac current of 50 Hz. The input power to the reactor to establish a stable gas discharge varied from 9.6 to maximum 15.3 watts (w). The effects of ZSM5, Fe–ZSM5, and Ni–ZSM5 catalysts combined with corona discharge for conversion of methane to more valued products have been addressed. It was found that in presence or absence of a catalyst in gas discharge reactor, the rate of methane and oxygen conversion increased upon higher input power supplied to the reactor. The effect of Fe–ZSM5 catalyst combined with gas discharge plasma yields C_2 hydrocarbons up to 21.9%, which is the highest productions of C_2 hydrocarbons in this work. The effect of combined Ni–ZSM5 and gas discharge plasma was mainly production of synthesis gas. The advantage of introducing ZSM5 to the plasma zone was increase in synthesis gas and acetylene production. The highest energy efficiency was 0.22 mmol/kJ, which belongs to lower rate of energy injection to the reactor.

INTRODUCTION

Although methane is an excellent raw material for production of fuels and chemicals, still its main use is as fuel for power generation and for domestic and industrial use. In many respects, methane is an ideal fuel for these purposes because of its availability in most populated centres, its ease of purification to remove sulphur compounds, and the fact that among the hydrocarbons, it has the largest heat of combustion relative to the amount of CO_2 formed. On the other hand, methane is a greatly underutilized resource for chemicals and liquid fuels [1]. Methane can be converted to chemicals and fuels in two ways, either via synthesis gas or directly into C_2 hydrocarbons or methanol. Almost all commercial processes for large-scale natural gas conversion involve synthesis gas production. Steam reforming is the dominant process for production of synthesis gas [2, 3] at high input energy as is shown in

$$CH_4 + H_2O \longleftrightarrow CO + 3H_2$$
$$\Delta H^0\ 298\ K = 206\ kJ/mol$$

1

The synthesis gas (CO + H_2) is converted to higher hydrocarbons or fuels via Fischer-Tropsch (FT) synthesis. The catalysts employed are based mainly on cobalt and iron at pressures as high as 22 bars and temperatures of higher than 560 K [4, 5].

To reduce energy costs it is possible to use electric power in a plasma process instead of thermal processing. In most of gas discharge plasma, free molecular radicals generated by excitation, dissociation, and ionization of gas molecules are essential for the subsequent free radical reactions. Control of electron energy by suitable design of discharge reactor and mode of gas discharge may lead to favorable products. Non-thermal plasmas at ambient temperature and pressure recently are being investigated as the promising alternative to convert methane to C_2 hydrocarbons [6–15].

In gas discharge plasma the main electrical energy is transferred to energetic electrons and active radical species rather than simple gas heating. As electrons possess minor mass compared to heavy ions they gain much higher speed in an electric field consequently they are in higher temperature state too. The electrons in a gas discharge collide with gas molecules and impart whole or a portion of their kinetic energy to exit the molecules to a higher energy states. Ultimately the molecule of the gas dissociates to radicals and other species and gives rise to synthesis of new products [16].

Recently zeolites have been found to be active for low temperature of methane conversion to higher hydrocarbon. Investigation of metal-substitute H-ZSM5 catalysts has also been reported [6]. In general low temperature methane conversion over zeolites has progressed in two directions: one is the modification of the performance of the zeolite as the support and the other is the modification of supported metal properties. Some mechanistic analyses have been presented [17, 18]. Relevant to these studies are the properties of zeolites related to their specific electronic structure. It is possible that the catalytic properties of a zeolite might be altered if it were electrically charged. No higher hydrocarbon activity was observed over this catalyst in the absence of corona discharge at any temperature up to 800°C. Charged catalytic activity obtained by the interaction of a corona discharge with a zeolite catalyst has led to low temperature methane conversion.

Due to its unique shape selectivity, solid acidity, ion exchangeability, pore size, thermal stability and structural network, ZSM-5 has been widely used as catalysts and sorbents in petroleum and petrochemical industry.

In this study methane conversion to olefins and synthesis gas using parallel plate discharge plasma is investigated. The experiments were performed over H-ZSM5 and its Fe and Ni substitute promoted catalyst. The effects of input power on methane and oxygen conversion and products selectivity are studied. Ultimately the yield of higher hydrocarbons and energy efficiency of the present work are evaluated.

EXPERIMENTAL

The schematic experimental setup used in this research is shown in Figure 1. The reactor was a quartz tube of 7 mm outer diameter (O.D) and 0.25 mm wall thickness. The electrodes are two similar stainless steel circular discs with 6.5 mm outer diameter and 2 mm thickness. The bottom electrode was kept at the potential of zero that is, was grounded and the upper electrode at different potentials. In all experiments performed methane and oxygen were mixed with helium as it helps the stability of gas discharge. In higher applied voltages to the reactor there were some liquid products collected in condenser which was kept it away without analysis.

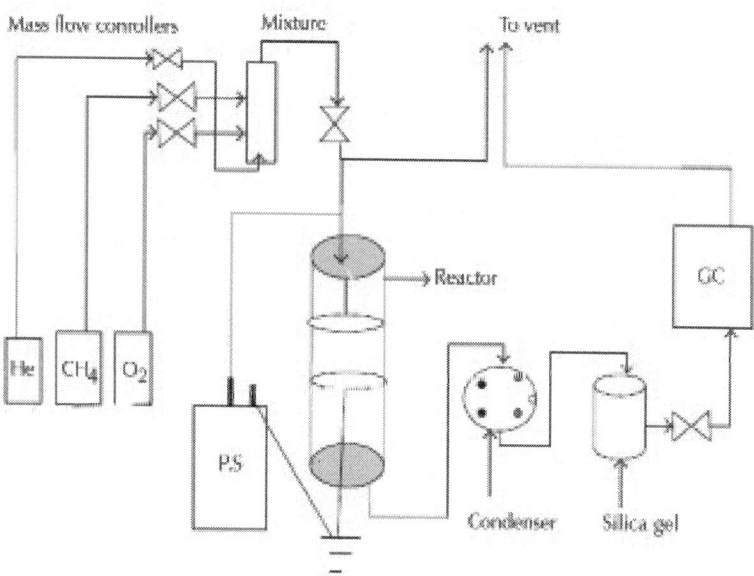

Figure 1: Platform of Corona Discharge with Feed and Analysis systems.

CATALYST PREPARATION

Fe-ZSM5 was prepared by solid state ion exchange method, H-ZSM5 (zeolyst Intel. with Si/Al = 50) was mixed with $FeCl_2$ in a ball mill for 75 minutes. The zeolite to iron chloride weight ratio was 2:1. The calcinations of the resulting powder was performed at 560°C for 5 hours, washed with De-ionized water to remove the anions and dried at 100°C for 16 h.

The ZSM-5 supported Ni catalysts were prepared by incipient wetness impregnation method using aqueous nitrates solution (Aldrich) as Ni precursors. Nickel nitrates, H-ZSM5 and water were thoroughly mixed at 60°C for 3 hours, dried at 120°C for 6 hours, followed by calcinations at 600°C during 5 hours in oxygen rich gas conditions. Reduction of catalyst was done by flowing hydrogen at 650°C for 1 hour.

The calcinations of H-ZSM5 were completed at 560°C in a tubular furnace for 6 hours in a flow of air. Table 1 represents the physical characteristics of the three catalysts used in the performed experiments.

Table 1: Textural properties of prepared catalysts

Catalyst	BET (m²/g)	Pore volume (cm³/g)	Si/Al	wt% metal
ZSM5	354	0.11	28	0
Fe-ZSM5	349	0.10	28	3
Ni-ZSM5	350	0.10	28	3

The following relations were used to determine the different parameters.

$$CH_4 \text{ conversion} = \left(\frac{\text{moles of } CH_4 \text{ consumed}}{\text{moles of } CH_4 \text{ introduced}}\right) \times 100$$

$$O_2 \text{ conversion} = \left(\frac{\text{moles of } O_2 \text{ consumed}}{\text{moles of } O_2 \text{ introduced}}\right) \times 100$$

$$\text{Selectivity of } C_2H_6 = 2 \times \left(\frac{\text{moles of } C_2H_6 \text{ formed}}{\text{moles of } CH_4 \text{ consumed}}\right) \times 100$$

$$\text{Selectivity of } C_2H_4 = 2 \times \left(\frac{\text{moles of } C_2H_4 \text{ formed}}{\text{moles of } CH_4 \text{ consumed}}\right) \times 100$$

$$\text{Selectivity of } C_2H_2 = 2 \times \left(\frac{\text{moles of } C_2H_2 \text{ formed}}{\text{moles of } CH_4 \text{ consumed}}\right) \times 100$$

$$\text{Yields of } C_2 \text{ hydrocarbons} = CH_4 \text{ conversion} \times \sum (\text{Selectivities of } C_2H_6, C_2H_4, C_2H_2)$$

$$\text{Energy Efficiency} = \frac{\text{Moles of } CH_4 \text{ Converted (mol/s)}}{\text{Energy Injected (w)}}$$

(2)

RESULTS AND DISCUSSION

The experiments were carried out in atmospheric pressure and room temperature keeping the total feed to the reactor for all experiments at 100 mL/minute in which the net methane flow rate was 16 mL/minute with methane to oxygen ratio as 4/1. First the feed was sent to the reactor under the plasma action only. In order to investigate the effect of prepared catalysts for converting of methane and oxygen, 0.45 gram of each catalyst H-ZSM5, Fe-ZSM5 and Ni-ZSM5 were separately loaded to the reactor. One of the catalysts was tested at a time, subsequently the reactor was cleaned and the next catalyst was loaded. For each of the catalysts an experiment was performed without the action of plasma, the analysis of results however indicate that there was no any new component in effluent gases from the reactor than the feed to the reactor up to 200°C. In all the experiments the reactor was never heated by external heaters but in some instances due to the conversion of a portion of input electrical energy to heat and release of heat due to exothermic reactions taking place in the plasma zone the wall of the reactor was at temperatures close to 150°C.

The conversions of methane and oxygen verses input power applied to the reactor are shown in Figures 2 and3. The conversion of CH_4 and O_2 increased upon increase in input of electrical power to the reactor. The Ni-ZSM5 shows the most significant effects in converting methane and oxygen as high as to 40% and 48.3% respectively. The effect of Fe-ZSM5 catalyst is interesting as the stable gas discharge plasma has been established at input power of 9.6 w, with methane conversion up to 31.4% and oxygen conversion to 36.8%, the stable gas discharge in presence of this catalyst ends up at input power of 11.3 w with methane and oxygen conversion of 36.8% and 44.6% respectively. The catalyst H-ZSM5 also shows enhancement in methane and oxygen conversion upon combined with gas discharge plasma. Presence of these catalysts had the effect of establishing a stable streamer discharge and it is believed that it was the main reason behind the increase of methane and oxygen conversions. The gas discharge shifts to an arc-like discharge more rapidly over Fe-ZSM5 catalyst compared to the other two tested catalysts.

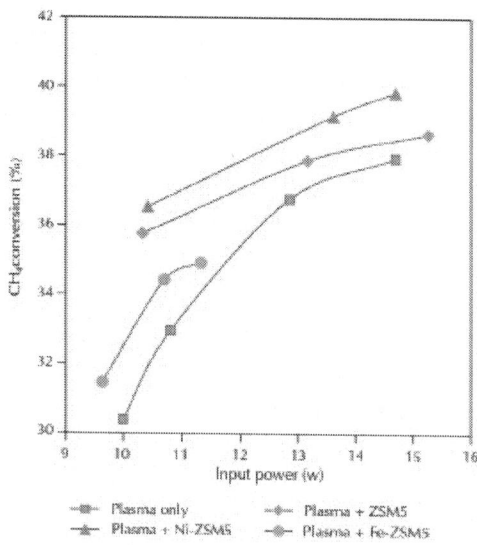

Figure 2: Methane conversion versus power consumed in the reactor.

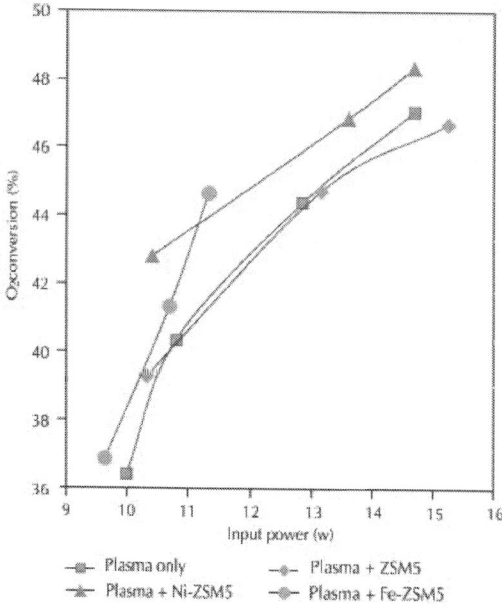

Figure 3: Oxygen conversion versus power consumed in the reactor.

The magnitude of input power is related with the strength of the electric field developed between the electrodes. By increasing the input power to the reactor the number of higher energy electrons that can ionize methane or oxygen molecules increases hence the conversion of CH_4 and O_2 increases.

Effect of Input Power to Selectivity of Products

The variations of products selectivity due to changes in electrical power consumed in the gas discharge reactor is shown in Figure 4. The selectivity of acetylene increases to about 16% at input power of 13 w; it reduces to 13.5% at power levels above 14 w. The highest selectivity to the most valuable hydrocarbon C_2H_4 is about 17.5% which belongs to input power of around 10 w, at higher power imparted to the reactor the selectivity to ethylene decreases. Selectivity to saturated hydrocarbon C_2H_6 is enhanced by applying higher input power to the reactor; it reaches to more than 18% at input powers around 14 w.

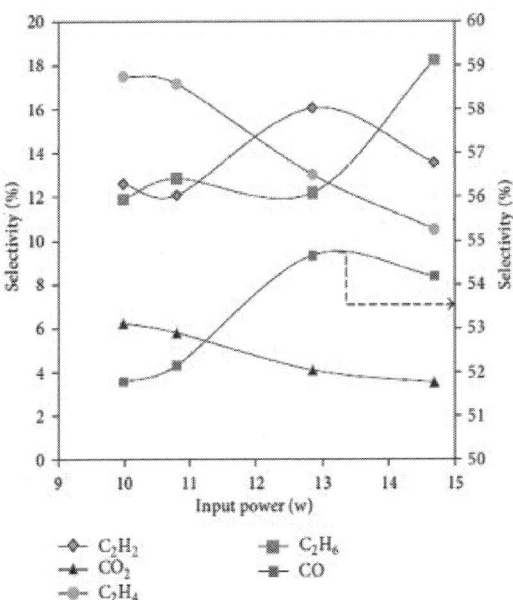

Figure 4: Selectivity of products versus input power to the reactor in corona discharge plasma.

The selectivity of CO is comparatively high and above 50% and doesn't change much with variations in input power to the reactor. As CO productions is always accompanied with H_2 gas according to the following equation:

$$CH_4 + \frac{1}{2}O_2 \longrightarrow CO + 2H_2 \qquad 3$$

Thus the whole process goes to produce C_2 hydrocarbons and synthsis gas (CO + H_2). The selectivity to unwanted carbon dioxide was however very low.

In another experiment, zeolite ZSM5 was brought into the reactor to investigate its combined effect with gas discharge plasma. Variations of products selectivity over this catalyst versus electrical power consumed in the reactor are shown in Figure 5. The main feature of this catalyst is high selectivity to acetylene, the selectivity of this component increases from 23% at 10.5 w to about 34.8% at 15 w. It can be concluded that the acidic nature of ZSM5 catalyst is the main source of methane dehydrogenation in plasma environment and hence increase in acetylene selectivity. Selectivity to ethane however is not remarkable; the highest selectivity to this product is about 9%. The selectivity to CO is 58.7 at 10.5 w; it reduces to 50% at input power of 15.2 w. Selectivity to unwanted CO_2 gas is however below 3%.

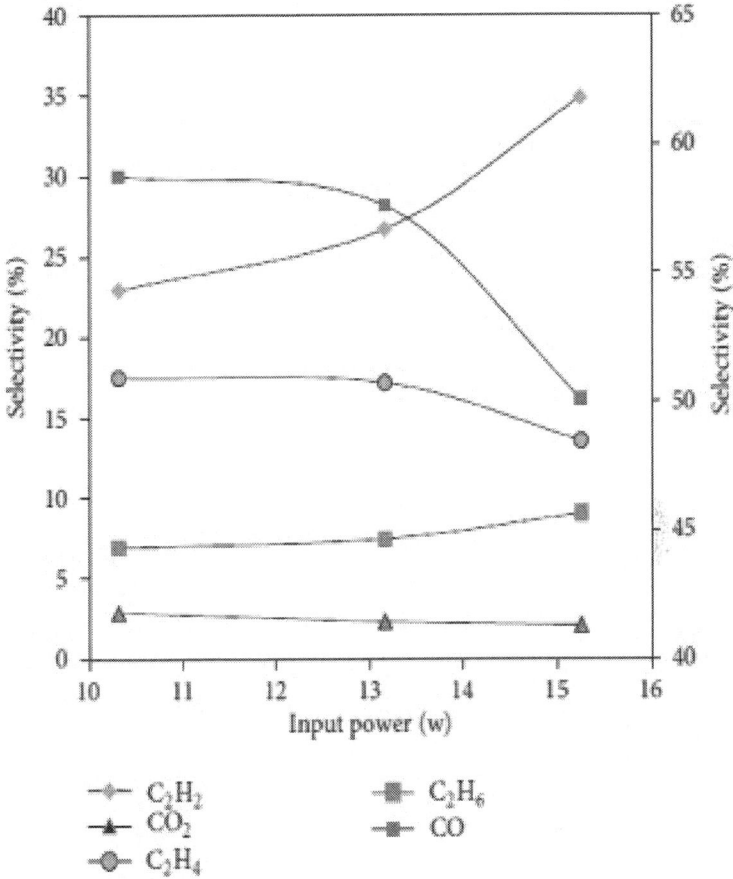

Figure 5: Selectivity of products versus input power supplied to the reactor over combined ZSM5 and corona discharge plasma.

In the next series of experiments Ni-ZSM5 catalyst was placed into the reactor to investigate its combined effect with gas discharge plasma. The variations in the products selectivity versus consumed power in the reactor are shown in Figure 6. The effect of this catalyst is enhanced in acetylene selectivity up to 24.4% and CO gas up to 66.2%.

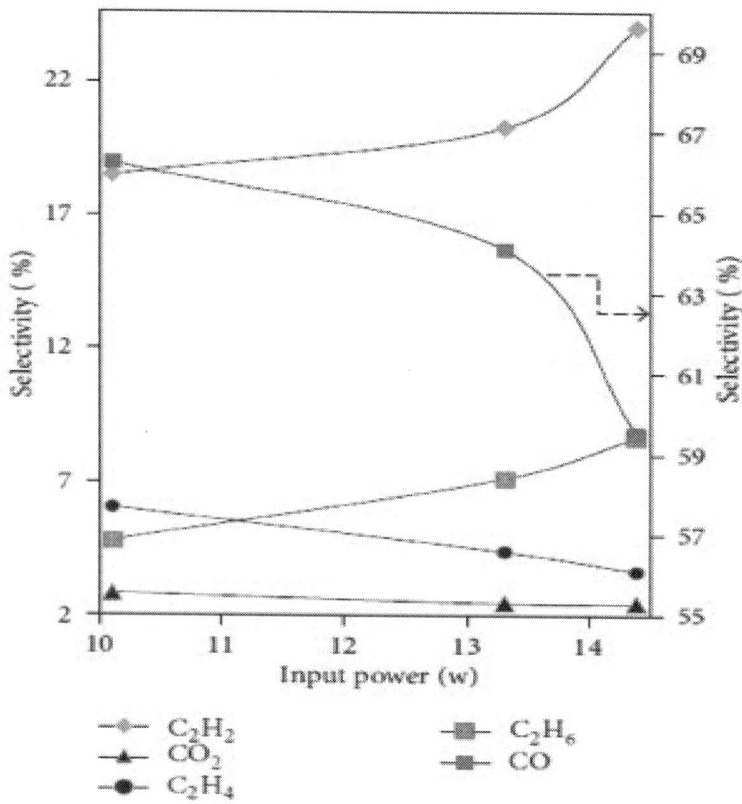

Figure 6: Selectivity of products versus power applied to the reactor over combined Ni-ZSM5 catalyst and gas discharge plasma.

When using Fe-ZSM5 as the catalyst combined with gas discharge plasma the results were noticeable. Figure 7 demonstrates at input power of 10.4 w, the selectivity to C_2H_2, C_2H_4 and C_2H_6 were 18.7%, 25.9% and 13.5% respectively. As the input power increases the selectivity to C_2H_4 continuously decreases to 19.4% at 14.7 w while the selectivity to C_2H_2 and C_2H_6 at the same input power is 21% and 19.4% respectively. The selectivity of CO is below 40% and production of synthesis gas was not dominant. The selectivity of CO_2 was however low and below 3%.

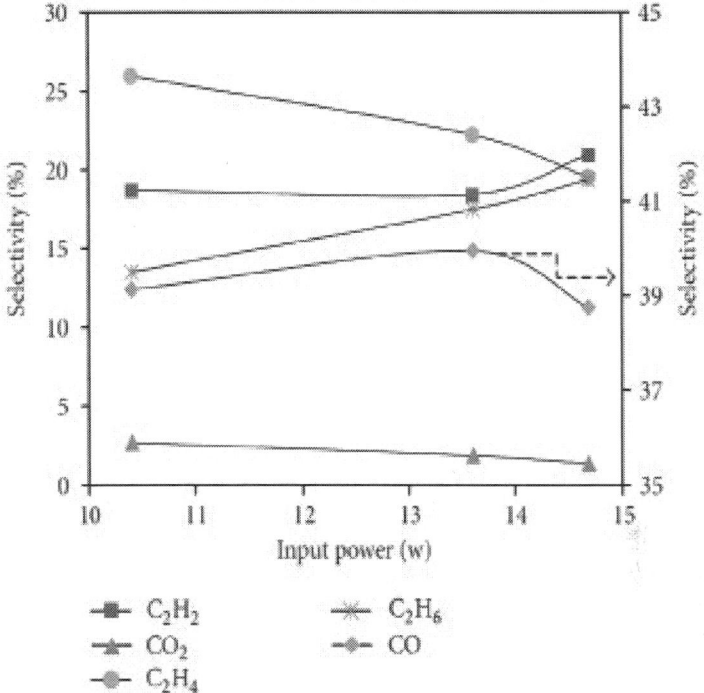

Figure 7: Selectivity of products versus power applied to the reactor over Fe-ZSM5 catalyst in plasma environment.

Yields of C_2 Hydrocarbons

The changes of total yield of C_2 hydrocarbons versus variations in power injected to the discharge reactor are shown in Figure 8. Compared with the gas discharge plasma as the only catalytic agent the ZSM5 has the effect of increasing the yield of C_2 hydrocarbons combined with the gas discharge plasma. In contrary the catalyst Ni-ZSM5 has effect of reducing yield of higher hydrocarbons compared with gas discharge plasma as the only catalytic effect. Introducing of Fe-ZSM5 to the plasma reactor is totally different this catalyst not only improves the yield of higher hydrocarbons to 20.3% at an input power of 9.6 w but also the stable gas discharge plasma initiates at lower levels of input power relative to other two catalyst and gas discharge without the presence of external catalyst.

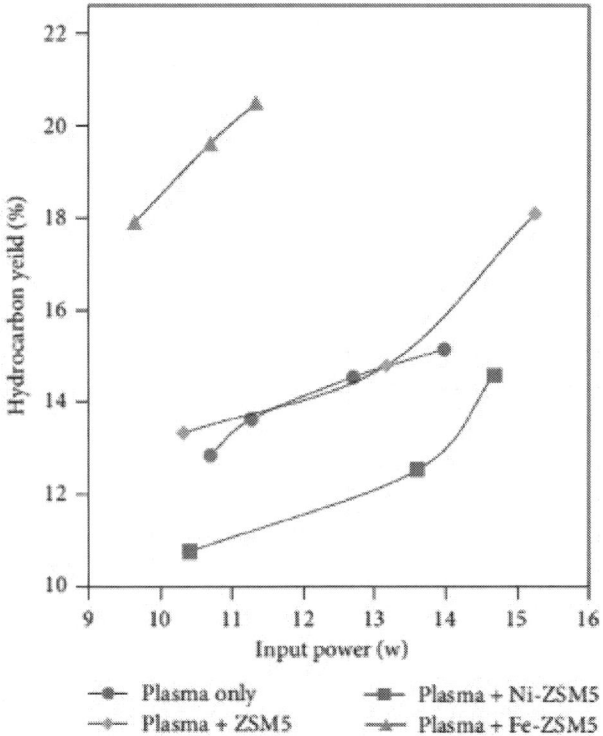

Figure 8: The yield of higher hydrocarbons versus input power supplied to the corona discharge reactor.

ENERGY EFFICIENCY

Figure 9 demonstrates the calculated energy efficiency of gas discharge plasma and three tested catalysts combined with gas discharge to convert one millimole of CH_4 to higher hydrocarbons and COx products. It is clear that the energy efficiency decreases at higher power imparted to the reactor irrespective of presence or absence of a catalyst in discharge reactor. The power supplied to the reactor at higher levels goes to heating up the bulk gas instead of increasing the kinetic energy of the electrons. The Fe-ZSM5 catalyst has reduced the values of energy efficiency in lower levels of power consumptions in the reactor but at the same time the extent of changes in the values of energy efficiency verses power to the reactor is limited to a small range. The Ni-ZSM also improves the energy efficiency to 0.22 mmol/kJ at input power of 10.4 w but the least energy efficiency for this catalyst is 0.17 at an input power of 14.7 w which stands after Fe-ZSM5 in higher input power to the reactor. The catalyst ZSM5 exhibits poor performance at higher input powers to the reactor as it shows 0.16 mmol/kJ 15.2 w. The energy efficiency of the gas discharge when there is no external catalyst in the reactor is high especially at lower levels of input power to the reactor. The energy efficiency for this catalyst is from 0.19 to 0.16 mmol/kJ for an input power of 10.8 to 14.7 w respectively.

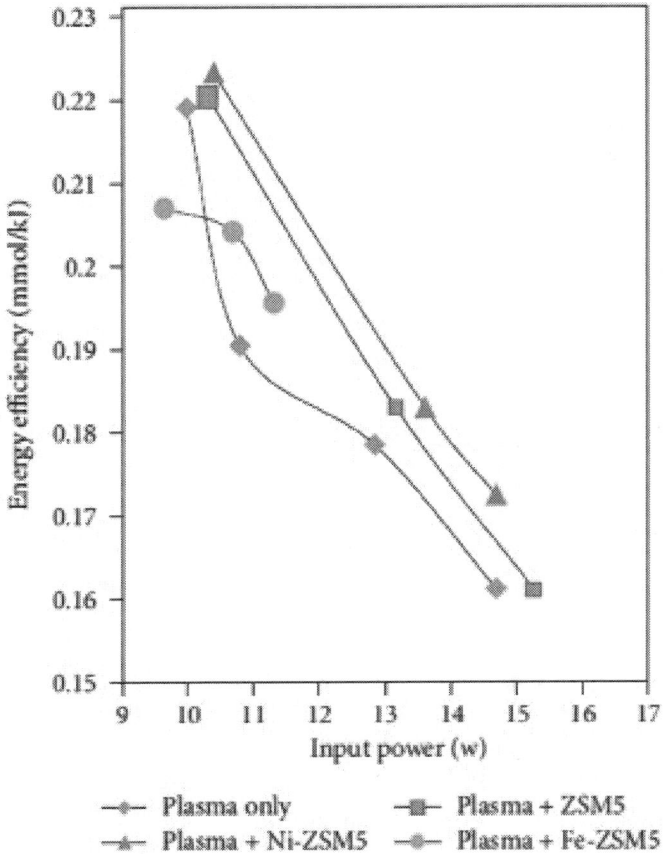

Figure 9: Energy efficiency versus input power supplied to the reactor.

Table 2 compares different cold plasma systems used to convert methane to more valued products [15].

Table 2: Energy efficiency in different plasma systems for conversion of methane

Plasma	Conversion rate (mmol/kJ)	Background gas
Microwave	0.22	CH_4
Silent	0.37	$CH_4 + O_2$
AC arc	0.15	$CH_4 + CO_2$
DC arc	3.1	$CH_4 + O_2$
Pulsed (240 pps)	0.36	$CH_4 + O_2$
Pulsed (8 kpps)	0.68	$CH_4 + CO_2$
Pulsed (8 kpps) reference study	1.2	CH_4
This study	0.22	$CH_4 + O_2$

CONCLUSIONS

Methane conversion which is the main constituent of natural gas to more valued products with ac corona discharge was investigated. The experiments were carried out in a quartz reactor with two parallel plate electrodes. In all experiments the rate of methane conversion was increased with the rate of electric power consumed in the reactor. The chemical inertness of alkens is generally overcome by the use of high temperatures and strongly acidic catalyst such as H- Zeolites or its metal exchanged catalysts. In this study instead of heating the zeolite based catalyst at high temperature the catalysts were placed in a strong electric field instead. The strong electric field changes the electrical properties of the surface of the catalyst and the conversion of methane takes place at temperatures below 200°C. The yield of C_2 hydrocarbons with Fe-ZSM5 was 20.9% and the highest energy efficiency in this study was 0.22 mmol/kJ while using a simple alternative current of 50 Hz. The ac corona discharge can be considered a promising method to convert methane to more valued products. It is a cheap and available technique and does not need very high and sophisticated technology to establish. Furthermore with the catalysts employed in this study it is possible to selectively convert methane to C_2 hydrocarbons or synthesis gas.

DISCLOSURE

This article is original and is not under consideration for publication elsewhere. In addition, the manuscript discusses about conversion of methane to more valued products which is across the spectrum of E-Journal of Chemistry. My co-worker in this research (Dr. Mohammad Irani) is aware of the submission and agrees to its publication.

REFERENCES

1. J. H. Lunsford, "Catalytic conversion of methane to more useful chemicals and fuels: a challenge for the 21st century," Catalysis Today, vol. 63, no. 2–4, pp. 165–174, 2000.
2. A. Holmen, "Direct conversion of methane to fuels and chemicals," Catalysis Today, vol. 142, no. 1-2, pp. 2–8, 2009.
3. J. R. Rostrup-Nielsen, "Production of synthesis gas," Catalysis Today, vol. 18, no. 4, pp. 305–324, 1993.
4. H. Schulz, "Short history and present trends of Fischer-Tropsch synthesis," Applied Catalysis A, vol. 186, no. 1-2, pp. 3–12, 1999.
5. A. Nakhaei Pour, M. R. Housaindokht, S. F. Tayyari, J. Zarkesh, and M. R. Alaei, "Deactivation studies of Fischer-Tropsch synthesis on nano-structured iron catalyst," Journal of Molecular Catalysis A, vol. 330, no. 1-2, pp. 112–120, 2010.
6. C. Liu, A. Marafee, R. Mallinson, and L. Lobban, "Methane conversion to higher hydrocarbons in a corona discharge over metal oxide catalysts with OH groups," Applied Catalysis A, vol. 164, no. 1-2, pp. 21–33, 1997.
7. C. J. Liu, R. Mallinson, and L. Lobban, "Nonoxidative methane

conversion to acetylene over zeolite in a low temperature plasma," Journal of Catalysis, vol. 179, no. 1, pp. 326–334, 1998.

8. A. Marafee, C. Liu, G. Xu, R. Mallinson, and L. Lobban, "An experimental study on the oxidative coupling of methane in adirect current corona discharge reactor over Sr/La 2 O 3 Catalyst," Industrial and Engineering Chemistry Research, vol. 36, no. 3, pp. 632–637, 1997.

9. L. M. Zhou, B. Xue, U. Kogelschatz, and B. Eliasson, "Partial oxidation of methane to methanol with oxygen or air in a nonequilibrium discharge plasma," Plasma Chemistry and Plasma Processing, vol. 18, no. 3, pp. 375–393, 1998.

10. C. J. Liu, R. Mallinson, and L. Lobban, "Comparative investigations on plasma catalytic methane conversion to higher hydrocarbons over zeolites," Applied Catalysis A, vol. 178, no. 1, pp. 17–27, 1999.

11. B. Eliasson, C.-J. Liu, and U. Kogelschatz, "Direct conversion of methane and carbon dioxide to higher hydrocarbons using catalytic dielectric-barrier discharges with zeolites," Industrial and Engineering Chemistry Research, vol. 39, no. 5, pp. 1221–1227, 2000.

12. F. M. Aghamir, N. S. Matin, A. H. Jalili, M. H. Esfarayeni, M. A. Khodagholi, and R. Ahmadi, "Conversion of methane to methanol in an ac dielectric barrier discharge," Plasma Sources Science and Technology, vol. 13, no. 4, pp. 707–711, 2004.

13. M. S. H. Tarverdi, Y. Mortazavi, A. A. Khodadadi, and S. Mohajerzadeh, "Synergetic effects of plasma, temperature and diluant on nonoxidative conversion of methane to C2+ hydrocarbons in a dielectric barrier discharge reactor," Iranian Journal of Chemistry and Chemical Engineering, vol. 24, no. 4, pp. 63–71, 2005.

14. S. L. Yao, E. Suzuki, N. Meng, and A. Nakayama, "Influence of rise time of pulse voltage on the pulsed plasma conversion of methane," Energy and Fuels, vol. 15, no. 5, pp. 1300–1303, 2001.

15. S. Yao, E. Suzuki, and A. Nakayama, "A novel pulsed plasma for chemical conversion," in Solid Films, vol. 390, no. 1-2, pp. 165–169, 2001.

16. C. Liu, A. Marafee, B. Hill, G. Xu, R. Mallinson, and L. Lobban, "Oxidative coupling of methane with ac and dc corona discharges," Industrial and Engineering Chemistry Research, vol. 35, no. 10, pp. 3295–3301, 1996.
17. V. R. Choudhary, A. K. Kinage, and T. V. Choudhary, "Lowtemperature nonoxidative activation of methane over H-galloaluminosilicate (MFI) zeolite," Science, vol. 275, no. 5304, pp. 1286–1288, 1997.
18. L. Y. Chen, L. W. Lin, Z. S. Xu, X. S. Li, and T. Zhang, "Dehydrooligomerization of methane to ethylene and aromatics over molybdenum/HZSM-5 catalyst," Journal of Catalysis, vol. 157, no. 1, pp. 190–200, 1995.

Chapter 2

n-Hexadecane Fuel for a Phosphoric Acid Direct Hydrocarbon Fuel Cell

Yuanchen Zhu[1,2], Travis Robinson[1], Amani Al-Othman[2,3], André Y. Tremblay[1], and Marten Ternan[4]

[1]Chemical and Biological Engineering, University of Ottawa, 161 Louis Pasteur, Ottawa, ON, Canada K1N 6N5
[2]Catalysis Centre for Research and Innovation, University of Ottawa, 30 Marie Curie, Ottawa, ON, Canada K1N 6N5
[3]Chemical Engineering, American University of Sharjah, Sharjah, UAE
[4]EnPross Incorporated, 147 Banning Road, Ottawa, ON, Canada K2L 1C5

ABSTRACT

The objective of this work was to examine fuel cells as a possible alternative to the diesel fuel engines currently used in railway locomotives, thereby decreasing air emissions from the railway transportation sector. We have investigated the performance of a phosphoric acid fuel cell (PAFC) reactor, with n-hexadecane, $C_{16}H_{34}$ (a model compound for diesel fuel, cetane number = 100). This is the first extensive study reported in the literature in which n-hexadecane is used directly as the fuel. Measurements were made to obtain both polarization curves and time-on-stream results. Because deactivation was observed hydrogen polarization curves were measured before and after n-hexadecane experiments, to determine the extent of deactivation of the membrane electrode assembly (MEA). By feeding water-only (no fuel) to the fuel cell anode the deactivated MEAs could be regenerated. One set of fuel cell operating conditions that produced a steady-state was identified. Identification of steady-state conditions is significant because it demonstrates that stable fuel cell operation is technically feasible when operating a PAFC with n-hexadecane fuel.

INTRODUCTION

Fuel cells offer many advantages for the conversion of the chemical energy in a fuel into electrical energy. Fuel cell energy efficiencies can be greater than those of conventional combustion engines. For example, because Carnot heat engines are limited to the maximum temperature that their materials can withstand, their theoretical energy efficiency is close to 67%. In contrast, fuel cells do not have materials limitations and can have larger theoretical energy efficiencies. Often emissions from fuel cells are generally less than those from combustion engines. In some applications fuel cells are competing successfully with batteries in part because they can use fuel continuously whereas batteries stop providing electrical power as soon as their charge has been exhausted.

Fossil fuels are usually the lowest cost source of energy and that is not apt to change in the foreseeable future. Unfortunately emissions from fossil fuels have a negative effect on the earth's climate. Direct hydrocarbon fuel cells (DHFCs) can have theoretical energy efficiencies near 95%. Their large energy efficiencies mean that a smaller quantity of fuel is required and therefore they will emit fewer emissions and have a smaller impact on climate change than heat engines or the more technological advanced fuel cells that use hydrogen or methanol as their fuels.

The purpose of this work was to decrease both greenhouse gas emissions (CO_2, CH_4, and N_2O) and air contaminants (NO_x, CO, HC, and SO_x) by replacing locomotive diesel engines with fuel cell engines. n-Hexadecane (cetane number = 100) was used as a model compound to represent commercial diesel fuels. A phosphoric acid fuel cell was used because its temperature is high enough to ensure that the n-hexadecane would be in the vapour phase if an appropriate steam/n-hexadecane ratio is used. Therefore the existence of two liquid phases within the fuel cell could be avoided.

Direct hydrocarbon fuel cells have other advantages. DHFC systems have lower capital costs than other fuel cell systems because the fuel processing systems (steam reforming, etc.) for hydrogen and methanol fuels are not required. In addition, the infrastructure already exists for diesel fuel and other petroleum derived fuels. That is not the case for hydrogen or methanol fuels. Storage of liquid fuels, such as diesel fuel, is much easier than storage of gaseous fuels such as hydrogen.

Unfortunately DHFCs have one major disadvantage. Their current densities are much smaller than those of hydrogen and methanol fuel cells. Work in our laboratory is being performed to understand the characteristics of DHFCs with a long-term objective of improving their performance.

William Grove demonstrated the first fuel cell operation in 1839 using hydrogen as the fuel. He was also credited with suggesting possible commercial opportunities if coal, wood, or other combustibles could replace hydrogen [1] which would be DHFCs. Direct hydrocarbon fuel cells were investigated intensely in the 1960s. Three reviews of the DHFC work up to that time are available [2–4].

Research on DHFCs has continued. Low-temperature fuel cell studies (<100°C) were performed on methane by Bertholet [5] and on propane by Cheng et al. [6] and by Savadogo and Rodriguez Varela [7, 8]. Heo et al. [9] performed intermediate temperature fuel cell studies (100–300°C) using propane. A larger number of DHFC studies have been performed on solid oxide fuel cells. Studies using low molecular weight hydrocarbons from methane to butane were performed by Steele et al. [10], Murray et al. [11], Zhu et al. [12], Gross et al. [13], and Lee et al. [14]. Larger molecules were studied by Ding et al. [15] (octane), Kishimoto et al. [16] (n-dodecane), and Zhou et al. [17] (jet fuel). Our own work has focused on modeling the fuel cell reactor [18–20], modeling the fuel cell catalyst [21–23], experimental development of an electrolyte that is appropriate for temperatures above the boiling point of water [24–26], and experimental fuel cell studies [27, 28].

Phosphoric acid fuel cell systems have an extensive development history. A 250–400 kW fuel cell system to produce stationary electric power was developed by Pratt and Whitney/ONSI/UTC Power. 300 units were built in 19 different countries. The company was sold to ClearEdge Power and was recently acquired by Doosan Industries. The phosphoric acid fuel cell technology has been documented extensively [29–32].

The fuel in this work was n-hexadecane. There were only three data points reported previously in a fuel cell study that examined a variety of fuels [33]. This is the first fuel cell study devoted exclusively to n-hexadecane. In a direct n-hexadecane phosphoric acid fuel cell, the overall reaction is

$$C_{16}H_{34}(g) + \frac{49}{2}O_2(g) \longrightarrow 16CO_2(g) + 17H_2O(g) \tag{1}$$

The anode half-cell reaction is

$$C_{16}H_{34}(g) + 32H_2O(g) \longrightarrow 16CO_2(g) + 98H^+ + 98e^- \tag{2}$$

The cathode half-cell reaction is

$$\frac{49}{2}O_2(g) + 98H^+ + 98e^- \longrightarrow 49H_2O(g) \tag{3}$$

where the (g) represents the gas phase. The anode stoichiometric ratio, SR = $H_2O/C_{16}H_{34}$, is 32. One mole of n-hexadecane reacts with 32 moles of water at the anode and generates 98 moles of protons and electrons. The protons migrate through the electrolyte to the cathode where the oxygen reduction reaction occurs.

Bagotzky et al. [34] described a reaction mechanism for direct hydrocarbon fuel cells using methane as a feedstock. The Bagotsky mechanism was modified, as shown in Figure 1, to describe n-hexadecane. The desired product is CO_2. However alcohols, aldehydes, carboxylic acids, and lower molecular weight hydrocarbons are possible by-products. Three reactions are shown in Figure 1: dehydrogenation (from both carbon and oxygen atoms), hydroxylation, and C–C bond cleavage. Two reactions are not shown: water dissociation ($H_2O \rightarrow H+OH$) and hydrogen atom ionization ($H \rightarrow H^+ + e^-$). Hydrogen ionization is an electrochemical reaction and therefore is influenced by potential. The other four reactions are chemical reactions and are not influenced by potential.

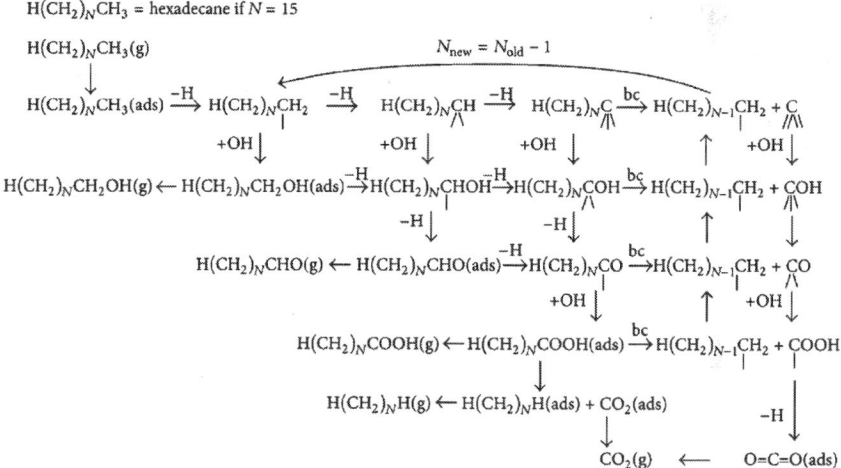

Figure 1: Diagram of a modified Bagotsky anode reaction mechanism: $H(CH_2)_NCH_3$ = hexadecane if N=15, +OH = Hydroxylation, −H = Dehydrogenation, and bc = C–C bond cleavage. Two reactions are not shown, water dissociation $H_2O = H + OH$ and the electrochemical reaction $H = H^+ + e^-$.

The objective of the work described here was to identify a set of operating conditions that would permit stable continuous operation of a direct hydrocarbon phosphoric acid fuel cell using n-hexadecane as the fuel.

EXPERIMENTAL

A schematic diagram of the direct n-hexadecane fueled phosphoric acid fuel cell (PAFC) system is shown in Figure 2. The overall system consists of an air cylinder, a hydrogen cylinder, one Galvanostat, two syringe pumps, a vaporizer, a phosphoric acid fuel cell (PAFC = Electrochem FC-25-02MA), and a fuel cell test station. Both gaseous and liquid fuels can be used in this fuel cell system. Deionized water and n-hexadecane were introduced into the vaporizer by the syringe pumps. The liquid fuels were expected to vaporize before reaching the anode of the fuel cell. Air was fed to the cathode at a constant flow rate. On those occasions when hydrogen was used as the fuel, the pumps were stopped and the valve in Figure 2 was opened.

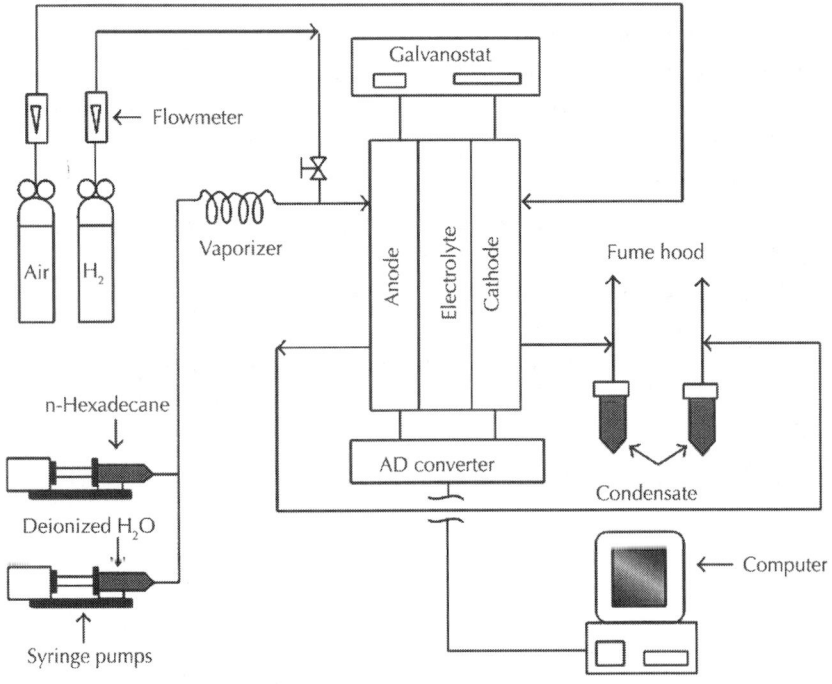

Figure 2: Diagram of a direct n-hexadecane fueled phosphoric acid fuel cell system.

The membrane electrode assembly (MEA) used in our fuel cell work had five layers: two gas diffusion layers (GDL), two catalyst layers (CL), and a liquid electrolyte layer. The gas diffusion layers were Teflon coated Toray paper.

The liquid electrolyte was initially 85% (14.6 M) phosphoric acid, which was held in a SiC matrix between the anode and cathode catalyst layers. Platinum (0.5 mg Pt/cm²) supported on carbon (10% Pt on C) was the catalyst in both anode and cathode catalyst layers. The fuel cells had a face area of 25 cm². A pin-type flow field was machined in a graphite plate. The current collectors were sheets of copper metal that had been gold plated on both sides. Silicone rubber flexible heaters were attached to each current collector.

Several types of experiments were performed. Hydrogen polarization curves were measured to determine the state of the MEA in the fuel cell. A polarization curve shows the potential difference as a function of current density. n-Hexadecane polarization curves were measured. Two types of time-on-stream experiments were performed (H_2O with n-$C_{16}H_{34}$ and H_2O only). The time-on-stream experiments were performed at (a) different molar ratios of water to n-hexadecane, (b) different current densities, and (c) different temperatures.

The following operating conditions were used. Separate syringe pumps were used to feed both water and n-hexadecane. The water flow rate was expressed as a function of the stoichiometric ratio (SR) of $H_2O/C_{16}H_{34}$ in (2) for the anode half reaction. A constant flow rate of n-hexadecane (0.2 mL/h) was used in all experiments. The two water flow rates and their stoichiometric ratios were 1 mL/h (2.5*SR, $H_2O/C_{16}H_{34}$ = 80) and 5.1 mL/h (12.9*SR, $H_2O/C_{16}H_{34}$ = 414). Some experiments were performed with only water being fed to the fuel cell. The experiments were performed at two temperatures, 160°C and 190°C.

A Hokuto Denko HA-301 Galvanostat was used to adjust the potential difference between the anode and cathode of the phosphoric acid fuel cell to maintain the chosen current at a constant value. The potential difference was recorded every second using a Lab View data logger.

RESULTS AND DISCUSSION

Two hydrogen/air polarization curves obtained with a PAFC are shown in Figure 3. The upper curve was the first experiment performed with a new MEA. The lower curve was measured after some conditioning experiments had been performed with low molecular weight hydrocarbons (ethylene, propane). It is an indication of the condition of the MEA at the beginning of this investigation and will be referred to as the Reference polarization curve. The open circuit potential in Figure 3 is about 0.93 V. It is comparable to the 0.9 V value reported by Fuller et al. [35] with an air cathode half-cell having a hydrogen Reference electrode.

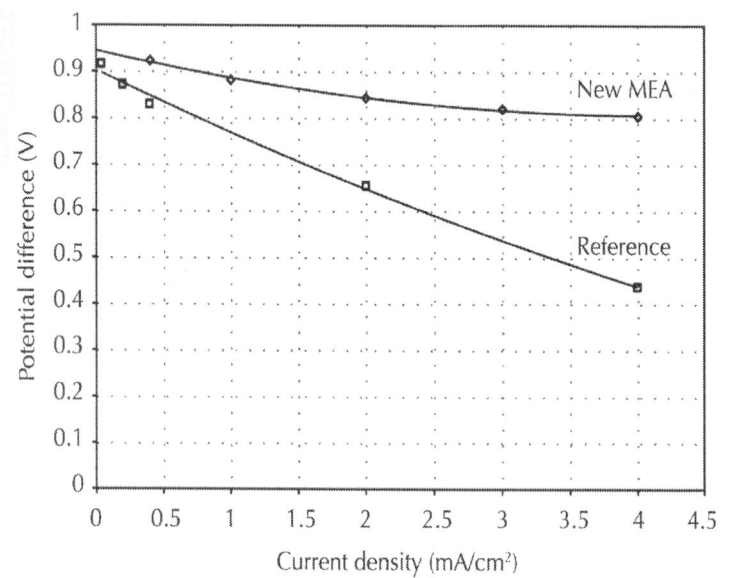

Figure 3: Polarization curve for a hydrogen fueled PAFC: potential difference between the electrodes (Volts) and current density (mA/cm^2). Open diamonds are data obtained on a new MEA. Open squares are data obtained on an MEA that had been conditioned in previous experiments with low molecular weight hydrocarbons (ethylene, propane). Anode: hydrogen flow rate = 9.6 mL/h. Cathode: air flow rate = 245 mL/min. Temperature = 160°C. Pressure = 1 atm.

The results of two time-on-stream experiments at 160°C are shown in Figure 4. Both curves show deactivation, indicated by a decrease in potential difference with time. The data show that deactivation continued for at least 20 hours. The two sets of data were obtained at different current densities and different $H_2O/n-C_{16}H_{34}$ molar ratios. The deactivation reported here with n-hexadecane is consistent with deactivation reported earlier by Okrent and Heath [36] during direct hydrocarbon fuel cell experiments with decane.

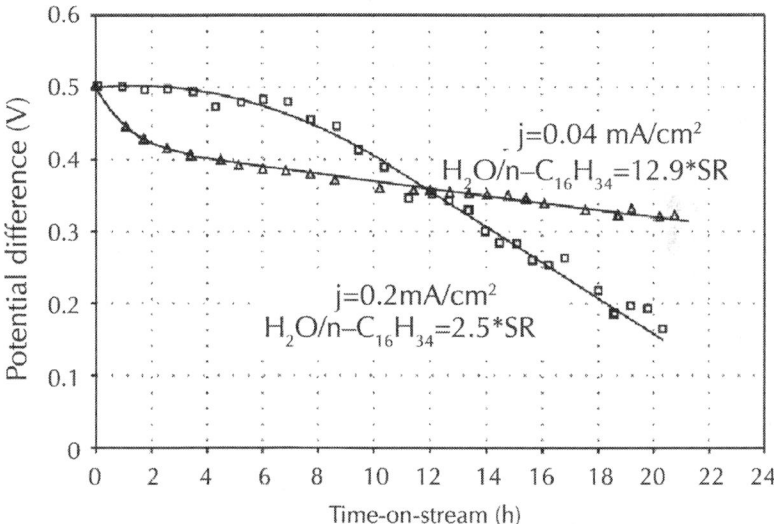

Figure 4: Potential difference between the electrodes (V) and time-on-stream (h) for an n-hexadecane fueled PAFC. Anode: n-hexadecane flow rate = 0.2 mL/h. Cathode: air flow rate = 245 mL/min. Temperature= 160°C. Pressure = 1 atm. Open triangles are data obtained with current density j=0.04 mA/cm², water flow rate = 5.1 mL/h, and $H_2O/n-C_{16}H_{34}$ =12.9*SR. Open squares are data obtained with current density j=0.2 mA/cm², water flow rate = 1 mL/h, and $H_2O/n-C_{16}H_{34}$ =2.5*SR.

Two hypotheses can be suggested to explain deactivation. Carbon monoxide, a reaction intermediate formed during the overall reaction to produce the CO_2, shown in Figure 1, could poison the platinum catalyst at the anode. Carbon monoxide is a well-known poison on fuel cell platinum catalysts [27]. The other possibility is the formation of carbonaceous deposits. Liebhafsky and Cairns [37] indicated the formation of dehydrogenated residues or carbonaceous materials during the operation of fuel cells with hydrocarbon fuels.

The current densities in Figure 4 were integrated with respect to time to obtain the cumulative amount of charge transferred. The potential difference in Figure 4 was plotted as a function of cumulative charge transferred in Figure 5. The data indicate that, at potential differences less than 0.4 V, the slopes of the two lines are the same. In other words, deactivation is a linear function of charge transferred. That observation suggests that deactivation, as represented by a decrease in potential difference, is related to some phenomenon that correlates with the amount of charge transferred, regardless of the $H_2O/n\text{-}C_{16}H_{34}$ molar ratio.

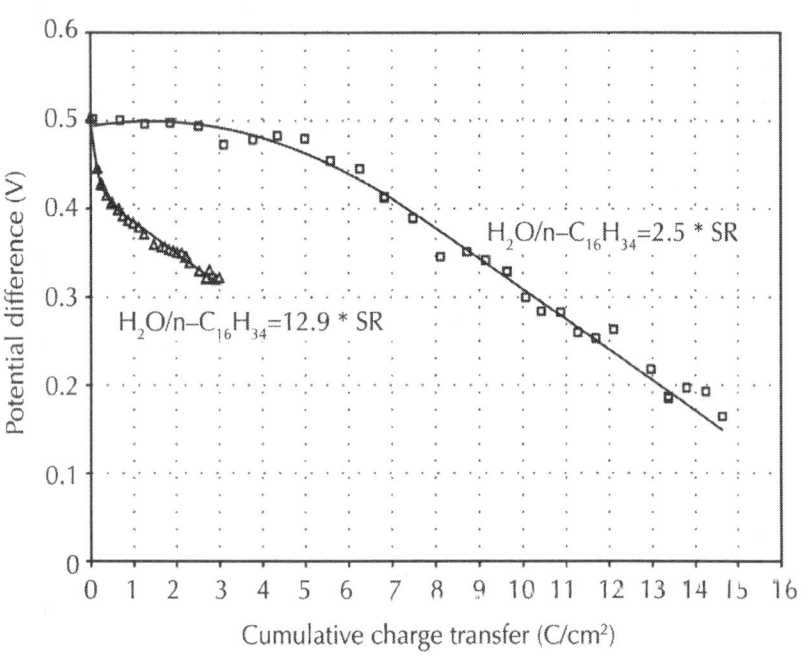

Figure 5: Potential difference between the electrodes (V) and cumulative charge transfer (C/cm²) for an n-hexadecane PAFC. Anode: n-hexadecane flow rate = 0.2 mL/h. Cathode: air flow rate = 245 mL/min. Pressure = 1 atm. Temperature = 160°C. Open triangles are data obtained with water flow rate = 5.1 mL/h, current density = 0.04 mA/cm², and $H_2O/n\text{-}C_{16}H_{34}$ =12.9*SR. Open squares are data obtained with water flow rate = 1 mL/h, current den-

sity = 0.2 mA/cm², and $H_2O/n-C_{16}H_{34}$ =2.5*SR .

A hydrogen/air polarization curve was measured using the PAFC after the first TOS experiment at 160°C (2.5*SR, $H_2O/n-C_{16}H_{34}$ = 81). In Figure 6, it is compared to the "Reference" hydrogen/air polarization curve, from Figure 3. The change between the Reference polarization curve and the one after the first time-on-stream experiment indicates that there had been a definite deterioration in the fuel cell performance. The data in Figure 6 are consistent with the deactivation observed during the TOS experiments in Figure 4. If the two polarization curves are compared at a constant value of potential difference, the current density is much smaller after the TOS measurements than before. Either the turnover frequency on a reaction site is much smaller or there are fewer reaction sites at which the reaction occurs. The only explanation is that something has prevented small hydrogen molecules from reacting to form electrons.

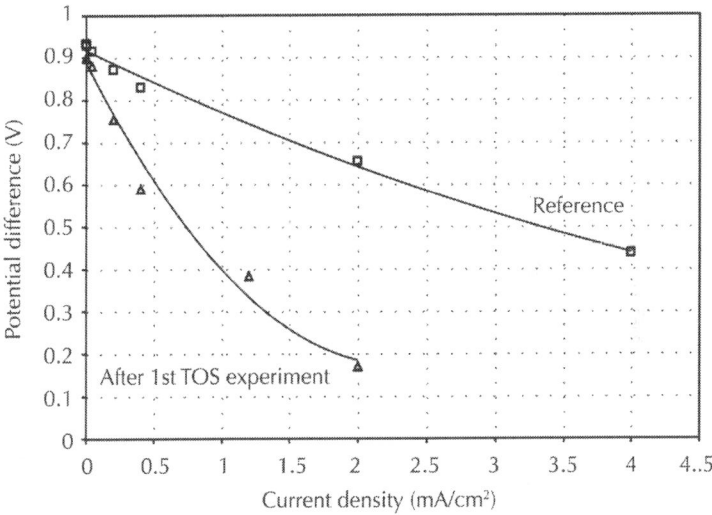

Figure 6: Polarization curve for a hydrogen fueled PAFC: potential difference between the electrodes and current density (mA/cm²). Anode: hydrogen flow rate = 9.6 mL/min. Cathode: air flow rate = 245 mL/min. Temperature = 160°C. Pressure = 1 atm. Open squares are the Reference polarization curve. Open triangles are the polarization curve obtained after the first time-on-stream experiment.

Since deactivation during the TOS experiments was observed using both sets of operating conditions at 160°C, further experiments were performed at a temperature of 190°C. The MEA was treated by operating sequentially with hydrogen (6 h), water (6 h), and hydrogen (6 h). Then a polarization curve was measured. The technique for measuring the polarization curve is indicated in Figure 7. The current density was set to a constant value. Then the potential difference was recorded until a steady-state value for the potential was obtained. For one datum point, corresponding to 0.4 mA/cm², the steady-state value of the potential difference was extrapolated from the data in Figure 7. Generally at least one hour was required to obtain a steady-state value for the potential difference. Finally, the steady-state values of the potential differences obtained in Figure 7 were used in Figure 8 to construct a polarization curve for the n-hexadecane/water-air fuel cell.

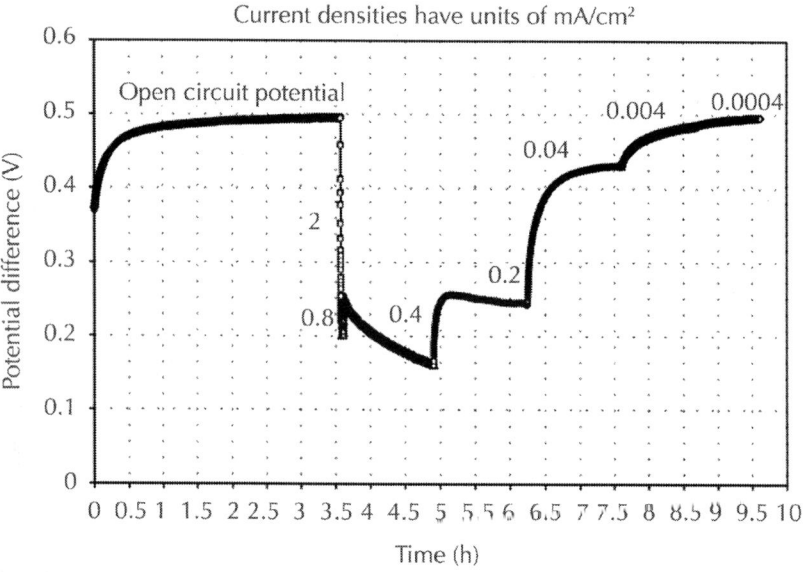

Figure 7: Potential difference between electrodes (V) as a function of time (h) obtained with a PAFC. Anode: water flow rate = 5.1 mL/h, n-hexadecane flow rate = 0.2 mL/h. Cathode: air flow rate = 245 mL/min. Temperature = 190°C, pressure = 1 atm, with an $H_2O/C_{16}H_{34}$ ratio =12.9*SR. The numbers on the top of each line represent different current densities.

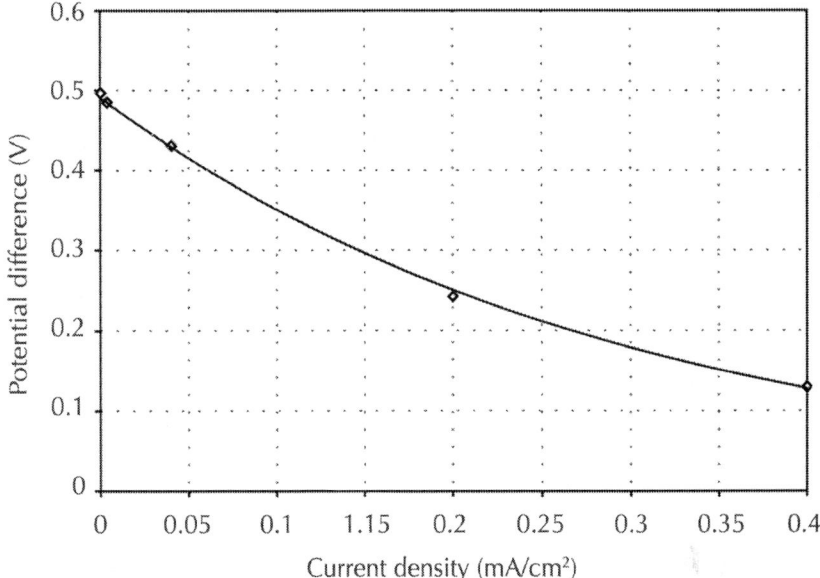

Figure 8: Polarization curve for an n-hexadecane fueled PAFC: potential difference between the electrodes (V) and current density (mA/cm²). Anode: water flow rate = 5.1 mL/h, n-hexadecane flow rate = 0.2 mL/h. Cathode: air flow rate = 245 mL/min. Temperature = 190°C. Pressure = 1 atm.

Some of the characteristics of the 190°C n-hexadecane/air polarization curve in Figure 8 are noteworthy. The open circuit potential of 0.5 V is much smaller than that of 0.93 V obtained for the hydrogen/air fuel cell in Figure 3. It suggests that the results in Figure 8 might represent the partial oxidation of carbon:

$$C + \frac{1}{2}O_2 = CO \qquad E^0_{298} = 0.711 \text{ V} \qquad (4)$$

as the rate limiting step in the overall reaction rather than the oxidation of n-hexadecane in (2). Equation (4) is composed of two half-cell reactions:

$$C + H_2O = CO + 2H^+ + 2e^- \quad \text{anode} \tag{5}$$

$$2H^+ + 2e^- + \frac{1}{2}O_2 = H_2O \quad \text{cathode} \tag{6}$$

The difference between 0.711 V and 0.5 V might be caused by a combination of factors: a temperature of 190°C rather than 25°C, a cathode oxygen mole fraction of 0.21, and an anode water vapour mole fraction representing equilibrium water vapour over phosphoric acid. The open circuit potential, 0.5 V, in Figure 8 is more consistent with the standard electrochemical potential of the partial oxidation of carbon to carbon monoxide reaction, 0.711 V, than with the standard electrochemical potential of the oxidation of carbon monoxide to carbon dioxide (CO + (1/2)O_2=CO_2, E^0_{298}=1.33V). Initially two possible hypotheses were suggested to explain deactivation: either carbon monoxide poisoning or deposition of carbonaceous material. Equation (4) is consistent with the carbonaceous material hypothesis and not consistent with carbon monoxide hypothesis. On that basis the hypothesis of deposition of carbonaceous material seems to be the most likely explanation for the deactivation observed during the time-on-stream experiments.

Time-on-stream measurements were also made at 190°C. The TOS results at 190°C are compared with those at 160°C in Figure 9. A steady-state operation was achieved for the last six hours of the experiment at 190°C. A steady-state operation is a highly desirable result that is not always achieved with a comparatively large hydrocarbon molecule, such as hexadecane. For example, Okrent and Heath [36] reported unsteady cycling during which both the potential and the current oscillated over time periods of approximately 15 minutes, when octane was the hydrocarbon fuel. Although we also observed cycling in some of our experiments, that phenomenon was not the object of our investigation. The fact that a steady-state has been demonstrated here for one set of operating conditions means that in principle fuel cells can operate continuously using n-hexadecane (and presumably other diesel type fuels).

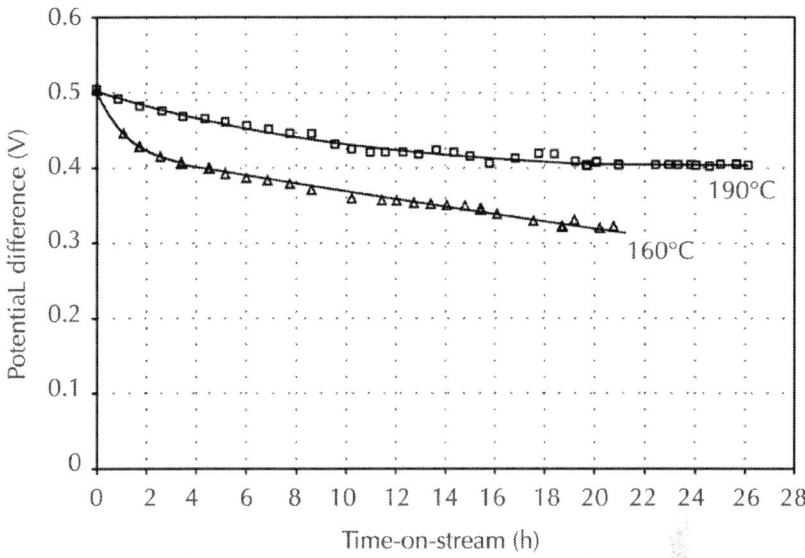

Figure 9: Potential differences between the electrodes (V) and time-on-stream (h) for an n-hexadecane PAFC. Current density j=0.04 mA/cm^2. Anode: water flow rate = 5.1 mL/h, n-hexadecane flow rate = 0.2 mL/h. Cathode: air flow rate = 245 mL/min. Pressure = 1 atm.

Cleaning the MEA with water was mentioned in the discussion pertaining to Figure 7. An example of water being the only reactant entering the fuel cell is shown in Figure 10. The data in Figure 10 were obtained from an MEA that had been used previously for 10 weeks in TOS experiments. When the current density was maintained constant at a value of 0.2 mA/cm^2 the potential difference decreased continuously for a period of 6 hours. That indicated that a progressively larger overpotential was necessary (a larger driving force was necessary) to maintain the current density at a constant value. When the current density was decreased to 0.1 mA/cm^2, there was an initial increase in the potential difference (smaller overpotential). The potential difference gradually decreased over the next 7 hours and then remained constant at 0.35 V for the last 6 hours.

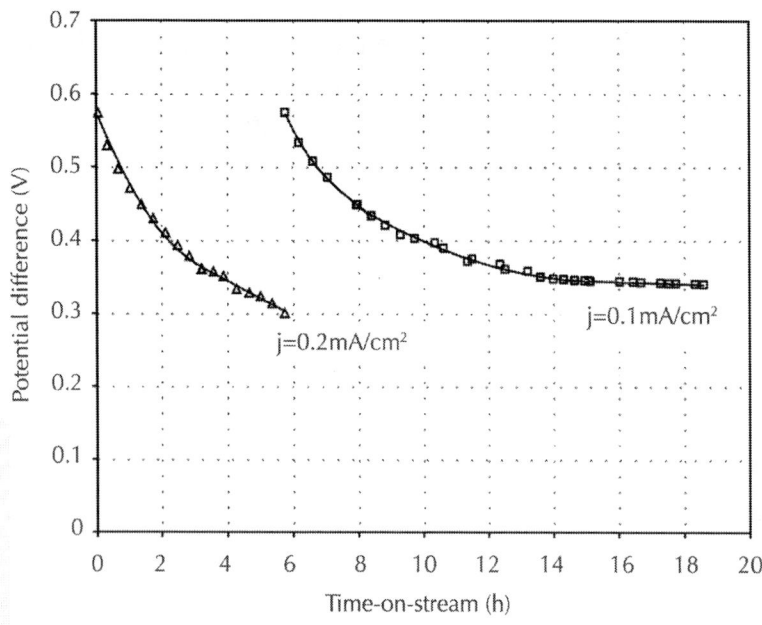

Figure 10: Potential differences between the electrodes (V) as a function of time-on-stream (h) when H_2O was the only feed stock for the anode of a PAFC with a fouled MEA. Anode: water flow rate = 5.1 mL/h. Cathode: air flow rate = 245 mL/min. Temperature = 190°C. Pressure = 1 atm.

The existence of a current density when only water was fed to the fuel cell would require that some reaction must have been occurring. Since no fuel (e.g., no n-hexadecane) was fed to the fuel cell, it is plausible that the reaction may have occurred between water and the carbonaceous material that had been previously deposited on the MEA. The existence of a current density would also require proton migration across the electrolyte. The occurrence of the anode reaction shown in (5) would be consistent with both of these requirements. The measurement of current density when only water was fed to the fuel cell is consistent with the hypothesis that carbonaceous material was formed during deactivation and was available for reaction during the water-only experiment.

After the water-only experiments in Figure 10 were completed, a hydrogen polarization curve was measured. It is compared with the Reference hydrogen polarization curve in Figure 11. A comparison of the results in Figure 11 (after the water-only experiment) with the results in Figure 6 (after the first TOS experiment) indicates that a substantial improvement was caused by the water-only treatment. That suggests the water-only experiment cleaned the MEA. Cleaning of the MEA would be consistent with removal of a carbonaceous deposit from the catalyst surface.

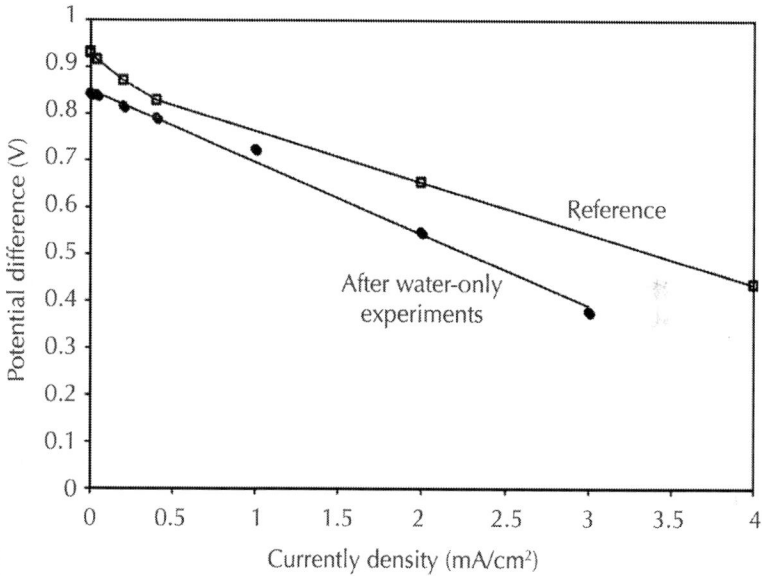

Figure 11: Polarization curve for a hydrogen fueled PAFC: potential difference between the electrodes and current density (mA/cm²). Anode: hydrogen flow rate = 9.6 mL/min. Cathode: air flow rate = 245 mL/min. Temperature = 160°C. Pressure = 1 atm. Solid circles are data obtained after a one-week experiment with water-only (on an MEA that had been used in TOS experiments for ten weeks). Open squares are data for the Reference polarization curve.

The results reported here can be compared with other fuel cell systems. Two of the important criteria are capital cost and energy efficiency. The capital cost is strongly influenced by the size of the fuel cell stack that in turn is a function of current density. The theoretical energy efficiency is related to the thermodynamic efficiency of the reactions that occur. The reaction networks that occur vary with the particular type of fuel cell system. The operating cost of a fuel cell system is strongly influenced by the energy efficiency.

Small current densities were obtained for the low-temperature (190°C) direct hydrocarbon (n-hexadecane) PAFC results without a reforming unit reported here. There are extensive reviews describing results obtained by PAFCs operating on hydrocarbons without a reforming unit [2–4]. In general the current densities are quite small. Therefore large reactors having a large capital cost would be required. In contrast Kim et al. [38] reported much larger current densities using a higher temperature (700°C) solid oxide fuel cell, SOFC, without a reforming unit when it was operating on synthetic diesel fuel. Interest in these systems, specifically the development of anodes, continues to be an active area of research [39, 40]. There have been several reviews of direct hydrocarbon SOFCs without reforming units [13, 41, 42]. Although no reforming unit was used, they indicate that internal reforming occurs [13]. Unfortunately the reforming reaction (internal or external) has a negative effect on energy efficiency. Approximately 25% of the hydrocarbon fuel must be used to provide the endothermic heat of reaction for the reforming reaction. At the low temperatures used in this study the reforming reaction is thermodynamically unfavourable and does not occur. Therefore the high-temperature SOFC systems will have a capital cost advantage over the lower temperature PAFC system used here. However, the lower temperature PAFC system used here will have a theoretical energy efficiency advantage over the SOFC system.

The use of an external reformer in combination with a PAFC system is a well-established technology that converts the hydrocarbon to hydrogen in a fuel processing system and then uses the hydrogen as the fuel in a fuel cell system. By the year 2006 more than 200 commercial plants had been sold [43]. Nevertheless research on improving the reforming process continues [44]. The reforming reaction in an external reformer has the same negative effect on energy efficiency that was mentioned above for internal reforming. The fuel processing system includes four processes: steam reforming, high-temperature water shift, low-temperature shift, and hydrogen purification. Equipment for those four processes has a substantial capital cost. In contrast there is no capital cost for a reformer/fuel processor with the low-temperature direct hydrocarbon PAFC system described here.

CONCLUSIONS

This study reported the first polarization curve ever measured for which n-hexadecane was the fuel at the anode of a fuel cell. The current densities were found to be very small.

Deactivation was observed in time-on-stream experiments. Deactivation, as measured by the change in potential difference, was found to be a linear function of the cumulative charge transferred across the electrolyte of the fuel cell. Deactivation during fuel cell experiments with n-hexadecane was confirmed by comparing hydrogen polarization curves before and after the time-on-stream measurements. For a given potential difference the current densities were much smaller for the hydrogen polarization curves measured after the time-on-stream experiments.

Experiments were performed in which water was the only reactant entering the fuel cell that had been used previously for 10 weeks in time-on-stream experiments. Current densities were measured during those experiments, indicating that the water must have reacted with some type of species that remained on the fuel cell catalyst at the end of the time-on-stream experiments. When a hydrogen polarization curve was measured at the end of the water-only experiments, it was close to that measured before the time-on-stream experiments. That indicates that the deactivating species on the surface of the platinum particles had been removed and that it was possible to regenerate deactivated MEAs.

A hypothesis that carbonaceous material was deposited on the platinum anode catalyst particles was suggested to explain the deactivation. Four types of observations were consistent with that hypothesis: (a) the change in potential difference during time-on-stream measurements; (b) when hydrogen polarization curves measured before and after the time-on-stream experiments were compared, the current densities measured after TOS were much smaller than those measured before TOS; (c) current densities were measured when water was the only reactant entering the fuel cell. In order to produce a current density, water must have reacted with some type of species that had been deposited on the surface of the platinum particles; and (d) the open circuit potential of a n-hexadecane fuel cell, 0.5 V, was much closer to the standard electrochemical potential for the carbon-water reaction, 0.711 V, than to that for the carbon monoxide-water reaction, 1.33 V. Observation (d) makes a hypothesis of deactivation by carbonaceous materials more likely than deactivation by carbon monoxide poisoning.

Steady-state operation of the n-hexadecane fuel cell, without additional deactivation, was observed at one set of fuel cell operating conditions. That observation demonstrates that stable fuel cell operation is technically feasible when n-hexadecane is the fuel at the anode of a fuel cell. It suggests there is merit in investigating fuel cell operation with commercial fuels such as petroleum diesel or biodiesel.

ACKNOWLEDGMENTS

The authors gratefully acknowledge that this research and development project was supported by a grant from Transport Canada's Clean Rail Academic Grant Program and by a Discovery Grant from the Canadian Government's Natural Sciences and Engineering Research Council.

REFERENCES

1. J. A. A. Ketelaar, "History," in Fuel Cell Systems, L. J. M. J. Blomen and M. N. Mugerwa, Eds., p. 20, Plenum Press, New York, NY, USA, 1993.
2. E. J. Cairns, "Anodic oxidation of hydrocarbons and the hydrocarbon fuel cell," Advances in Electrochemistry Science and Electrochemical Engineering, vol. 8, pp. 337–392, 1971.
3. J. O. M. Bockris and S. Srinivasan, "Electrochemical combustion of organic substances," in Fuel Cells: Their Electrochemistry, pp. 357–411, McGraw-Hill, New York, NY, USA, 1969.
4. H. A. Liebhafsky and E. J. Cairns, "The direct hydrocarbon fuel cell with aqueous electrolytes," in Fuel Cells and Fuel Batteries, pp. 458–523, John Wiley & Sons, New York, NY, USA, 1968.
5. S. Bertholet, Oxydation Electrocatalytique du Methane [Ph.D. Dissertation], Université de Poitiers, Poitiers, France, 1998.
6. C. K. Cheng, J. L. Luo, K. T. Chuang, and A. R. Sanger, "Propane fuel cells using phosphoric-acid-doped polybenzimidazole

7. O. Savadogo and F. J. Rodriguez Varela, "Low-temperature direct propane polymer electrolyte membranes fuel cell," Journal of New Materials for Electrochemical Systems, vol. 4, no. 2, pp. 93–97, 2001.

8. F. J. Rodríguez Varela and O. Savadogo, "Real-time mass spectrometric analysis of the anode exhaust gases of a direct propane fuel cell," Journal of the Electrochemical Society, vol. 152, no. 9, pp. A1755–A1762, 2005.

9. P. Heo, K. Ito, A. Tomita, and T. Hibino, "A proton-conducting fuel cell operating with hydrocarbon fuels," Angewandte Chemie—International Edition, vol. 47, no. 41, pp. 7841–7844, 2008.

10. B. C. H. Steele, I. Kelly, H. Middleton, and R. Rudkin, "Oxidation of methane in solid state electrochemical reactors," Solid State Ionics, vol. 28–30, no. 2, pp. 1547–1552, 1988.

11. E. P. Murray, T. Tsai, and S. A. Barnett, "A direct-methane fuel cell with a ceria-based anode," Nature, vol. 400, no. 6745, pp. 649–651, 1999.

12. W. Zhu, C. Xia, J. Fan, R. Peng, and G. Meng, "Ceria coated Ni as anodes for direct utilization of methane in low-temperature solid oxide fuel cells," Journal of Power Sources, vol. 160, no. 2, pp. 897–902, 2006.

13. M. D. Gross, J. M. Vohs, and R. J. Gorte, "Recent progress in SOFC anodes for direct utilization of hydrocarbons," Journal of Materials Chemistry, vol. 17, no. 30, pp. 3071–3077, 2007.

14. J. G. Lee, C. M. Lee, M. Park, and Y. G. Shul, "Direct methane fuel cell with $La_2Sn_2O_7$–Ni–$Gd_{0.1}Ce_{0.9}O_{1.95}$ anode and electrospun $La_{0.6}Sr_{0.4}Co_{0.2}Fe_{0.8}O_{3-\delta}$–$Gd_{0.1}Ce_{0.9}O_{1.95}$ cathode," Royal Society of Chemistry Advances, vol. 3, no. 29, pp. 11816–11822, 2013.

15. D. Ding, Z. Liu, L. Li, and C. Xia, "An octane-fueled low temperature solid oxide fuel cell with Ru-free anodes," Electrochemistry Communications, vol. 10, no. 9, pp. 1295–1298, 2008.

16. H. Kishimoto, K. Yamaji, T. Horita et al., "Feasibility of liquid hydrocarbon fuels for SOFC with Ni-ScSZ anode," Journal of Power Sources, vol. 172, no. 1, pp. 67–71, 2007.

17. Z. F. Zhou, C. Gallo, M. B. Pague, H. Schobert, and S. N. Lvov, "Direct oxidation of jet fuels and Pennsylvania crude oil in a solid oxide fuel cell," Journal of Power Sources, vol. 133, no. 2, pp. 181–187, 2004.
18. G. Psofogiannakis, Y. Bourgault, B. E. Conway, and M. Ternan, "Mathematical model for a direct propane phosphoric acid fuel cell," Journal of Applied Electrochemistry, vol. 36, no. 1, pp. 115–130, 2006.
19. H. Khakdaman, Y. Bourgault, and M. Ternan, "Computational modeling of a direct propane fuel cell," Journal of Power Sources, vol. 196, no. 6, pp. 3186–3194, 2011.
20. H. R. Khakdaman, Y. Bourgault, and M. Ternan, "Direct propane fuel cell anode with interdigitated flow fields: two-dimensional model," Industrial & Engineering Chemistry Research, vol. 49, no. 3, pp. 1079–1085, 2010.
21. G. Psofogiannakis, A. St-Amant, and M. Ternan, "Ab-initio DFT study of methane electro-oxidation mechanism on platinum," Journal of Physical Chemistry B, vol. 110, pp. 24593–24605, 2006.
22. S. Vafaeyan, A. St-Amant, and M. Ternan, "Nickel alloy catalysts for the anode of a high temperature PEM direct propane fuel cell," Journal of Chemistry, vol. 2014, Article ID 151638, 8 pages, 2014. ·
23. S. Vafaeyan, A. St-Amant, and M. Ternan, "Propane fuel cells: selectivity for partial or complete reaction," Journal of Fuels, vol. 2014, Article ID 485045, 9 pages, 2014. ·
24. A. Al-Othman, A. Y. Tremblay, W. Pell et al., "A modified silicic acid (Si) and sulphuric acid (S)-ZrP/PTFE/glycerol composite membrane for high temperature direct hydrocarbon fuel cells," Journal of Power Sources, vol. 224, pp. 158–167, 2013.
25. A. Al-Othman, A. Y. Tremblay, W. Pell, Y. Liu, B. A. Peppley, and M. Ternan, "The effect of glycerol on the conductivity of Nafion-free ZrP/PTFE composite membrane electrolytes for direct hydrocarbon fuel cells," Journal of Power Sources, vol. 199, pp. 14–21, 2012.
26. A. Al-Othman, A. Y. Tremblay, W. Pell et al., "Zirconium phosphate as the proton conducting material in direct hydrocarbon polymer

electrolyte membrane fuel cells operating above the boiling point of water," Journal of Power Sources, vol. 195, no. 9, pp. 2520–2525, 2010. ··

27. C. G. Farrell, C. L. Gardner, and M. Ternan, "Experimental and modelling studies of CO poisoning in PEM fuel cells," Journal of Power Sources, vol. 171, no. 2, pp. 282–293, 2007.

28. R. Fonocho, C. L. Gardner, and M. Ternan, "A study of the electrochemical hydrogenation of o-xylene in a PEM hydrogenation reactor," Electrochimica Acta, vol. 75, pp. 171–178, 2012.

29. N. Sammes, R. Bove, and K. Stahl, "Phosphoric acid fuel cells: fundamentals and applications,"Current Opinion in Solid State and Materials Science, vol. 8, no. 5, pp. 372–378, 2004.

30. J. M. King and H. R. Kunz, "Phosphoric acid electrolyte fuel cells," in Handbook of Fuel Cells, W. Vielstich, A. Lamm, H. A. Gasteiger, and H. Yokokawa, Eds., vol. 1, pp. 287–300, John Wiley & Sons, New York, NY, USA, 2010.

31. S. R. Choudhury, "Phosphoric acid fuel cell technology," in Recent Trends in Fuel Cell Science and Technology, S. Basu, Ed., pp. 188–216, Springer, New York, NY, USA, 2007.

32. W. Grubb and C. J. Michalske, "A high performance propane fuel cell operating in the temperature range of 150°–200°C," Journal of The Electrochemical Society, vol. 111, no. 9, p. 1015, 1964. ·

33. H. A. Liebhafsky and W. T. Grubb, "Normal alkanes at platinum anodes," Fuel Preprints, vol. 11, no. 2, p. 134, 1967.

34. V. S. Bagotzky, Y. B. Vassiliev, and O. A. Khazova, "Generalized scheme of chemisorption, electrooxidation and electroreduction of simple organic compounds on platinum group metals,"Journal of Electroanalytical Chemistry and Interfacial Electrochemistry, vol. 81, no. 2, pp. 229–238, 1977.

35. T. F. Fuller, F. J. Luczak, and D. J. Wheeler, "Electrocatalyst utilization in phosphoric acid fuel cells,"Journal of the Electrochemical Society, vol. 142, no. 6, pp. 1752–1757, 1995.

36. E. H. Okrent and C. E. Heath, "A liquid hydrocarbon fuel cell battery," in Fuel Cell Systems, B. Baker, Ed., Advances in Chemistry, pp. 328–340, American Chemical Society, Washington, DC, USA, 1969.

37. H. A. Liebhafsky and E. J. Cairns, "The direct hydrocarbon fuel cell with aqueous electrolytes," in Fuel Cells and Fuel Batteries, pp. 485–510, John Wiley & Sons, New York, NY, USA, 1968.
38. H. Kim, S. Park, J. M. Vohs, and R. J. Gorte, "Direct oxidation of liquid fuels in a solid oxide fuel cell," Journal of the Electrochemical Society, vol. 148, no. 7, pp. A693–A695, 2001.
39. S. Islam and J. M. Hill, "Barium oxide promoted Ni/YSZ solid-oxide fuel cells for direct utilization of methane," Journal of Materials Chemistry A, vol. 2, no. 6, pp. 1922–1929, 2014. ··
40. C. Yang, J. Li, Y. Lin, J. Liu, F. Chen, and M. Liu, "In-situ fabrication of CoFe allot nanoparticles structure $(Pr_{0.4}Sr_{0.6})_3$ $(Fe_{0.85}Nb_{0.15})O_7$ ceramic anode for direct hydrocarbon solid oxde fuel cells," Nano Energy, vol. 11, pp. 704–711, 2015.
41. S. McIntosh and R. J. Gorte, "Direct hydrocarbon solid oxide fuel cells," Chemical Reviews, vol. 104, no. 10, pp. 4845–4865, 2004.
42. Y. Zhao, C. Xia, L. Jia et al., "Recent progress on solid oxide fuel cell: lowering temperature and utilizing non-hydrogen fuels," International Journal of Hydrogen Energy, vol. 38, no. 36, pp. 16498–16517, 2013.
43. S. Srinivasan, Fuel Cells: From Fundamentals to Applications, Springer, New York, NY, USA, 2006.
44. M. R. Walluk, J. Lin, M. G. Waller, D. F. Smith, and T. A. Trabold, "Diesel auto-thermal reforming for solid oxide fuel cell systems: anode off-gas recycle simulation," Applied Energy, vol. 130, pp. 94–102, 2014.

Chapter 3

Syngas Generation from Methane Using a Chemical-Looping Concept: A Review of Oxygen Carriers

Kongzhai Li[1,2], Hua Wang[1], and Yonggang We[1]

[1]Engineering Research Center of Metallurgical Energy Conservation and Emission Reduction, Kunming University of Science and Technology, Ministry of Education, Kunming, Yunnan 650093, China

[2]Faculty of Metallurgy and Energy Engineering, Kunming University of Science and Technology, Room 217, Kunming, Yunnan 650093, China

ABSTRACT

Conversion of methane to syngas using a chemical-looping concept is a novel method for syngas generation. This process is based on the transfer of gaseous oxygen source to fuel (e.g., methane) by means of a cycling process using solid oxides as oxygen carriers to avoid direct contact between fuel and gaseous oxygen. Syngas is produced through the gas-solid reaction between methane and solid oxides (oxygen carriers), and then the reduced oxygen carriers can be regenerated by a gaseous oxidant, such as air or water. The oxygen carrier is recycled between the two steps, and the syngas with a ratio of $H_2/CO = 2.0$ can be obtained successively. Air is used instead of pure oxygen allowing considerable cost savings, and the separation of fuel from the gaseous oxidant avoids the risk of explosion and the dilution of product gas with nitrogen. The design and elaboration of suitable oxygen carriers is a key issue to optimize this method. As one of the most interesting oxygen storage materials, ceria-based and perovskite oxides were paid much attention for this process. This paper briefly introduced the recent research progresses on the oxygen carriers used in the chemical-looping selective oxidation of methane (CLSOM) to syngas.

INTRODUCTION

Methane, the principal constituent of natural gas and coal-bed gas, is an excellent raw material for production of fuels and chemicals [1]. Conversion of methane to value-added products can be achieved in two ways, either via syngas (a mixture of CO and H_2) as an intermediate or directly into C_2 and higher hydrocarbons. Since the direct catalytic conversion of methane is inefficient, almost all the commercial processes for large scale chemical utilization of methane such as Fischer-Tropsch synthesis, methanol, or dimethyl ether production involve syngas [2].

Syngas generation from methane can be achieved in three routes: water steam reforming (SMR), carbon dioxide reforming (CDR), and partial oxidation of methane (POM) [3]. The two reforming reactions are all highly endothermic and operated at high temperature and high pressure, termed as costly chemical processes. POM technology, by contrast, is a mildly exothermic route, which makes the process less energy and capital cost than the reforming routes. In addition, it also allows excellent syngas yield in compact reactors due to the fast reaction rate and product selectivity [3–5]. However, this technology requires additional safety measures to avoid the risk of explosion due to the premixing of CH_4/O_2 mixture and pure oxygen supply to avoid the dilution of syngas by nitrogen and the formation of NO_x [6], which partly offset its advantages in the saving of energy and capital cost. To avoid such problems, a chemical-looping concept was proposed to use in the POM technology.

CHEMICAL-LOOPING CONCEPT

The term "chemical looping" is a new concept for fuels conversion, which is based on the transfer of oxygen from gaseous oxygen source to the fuel by means of a cycling process using solid oxides as oxygen carriers to avoid direct contact between fuel and gaseous oxygen [7]. In the case of methane as fuel, the schematic of the chemical-looping process was shown in Figure 1. Lattice oxygen in oxygen carriers was used to oxidize methane, and then the reduced oxygen carriers can be reoxidized by gaseous oxidant to restore its initial state. Two interconnected reactors or fluidized-bed system are used in this technique to achieve the circulation of oxygen carrier between the oxidizing and reducing steps.

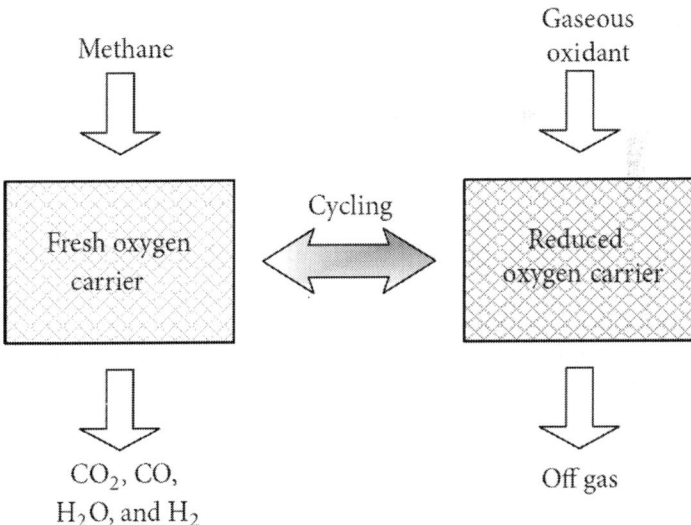

Figure 1: Schematic of chemical-looping concept for methane conversion.

The first design using this concept was developed for power generation, which is known with the general term "chemical looping combustion" (CLC). For this process, the oxygen carrier can convert fuels to H_2O and CO_2, and the reduced oxygen carriers must be reoxidized by air. Because of the separation of fuels from air, this technology is identified as owning inherent advantages for CO_2 separation with minimum energy losses [8]. Further designs of this concept were used in the syngas production from methane. After methane is oxidized to CO_2 and H_2O by oxygen carriers, the by-product gases (CO_2 and H_2O) were introduced into another reactor (reforming reactor) to reform with additional methane to produce syngas in the presence of a reforming catalyst (e.g., Ni/Al_2O_3) [9, 10]. Since the additional reforming process is a highly endothermic reaction needing large energy supply, this technology is less-than-ideal for syngas generation.

On the other hand, the direct generation of syngas by the reaction between oxygen carriers and methane is more acceptable, but this process needs an oxygen carrier owning ability to selectively oxidize methane. This vision was firstly realized over CeO_2 oxygen carrier, and the authors also proposed that the reduced oxygen carrier can be reoxidized by H_2O with obtaining H_2 simultaneously [11, 12]. In this case, the design and elaboration of suitable oxygen carriers with high activity, selectivity, and redox stability for methane selective oxidation is a key issue for this technology.

Comparing with the traditional POM process, the chemical-looping concept allows air instead of pure oxygen as oxygen source without the dilution of product gas with nitrogen, which brings about considerable cost saving. When using H_2O as an oxidant (two-step SRM process), it gives the possibility of coproduction of pure hydrogen without separating equipments and syngas with a H_2/CO ratio of 2.0 which is ideal for the major downstream processes such as methanol production or Fischer-Tropsch synthesis. The present paper would mainly discuss the progresses on the oxygen carriers for this technology.

OXYGEN CARRIERS FOR CHEMICAL-LOOPING SELECTIVE OXIDATION OF METHANE (CLSOM)

Chemical-looping concept involves the use of a redox cycle process of chosen oxygen carriers to implement the selective oxidation of methane to syngas. The yield of syngas depends on the activity and selectivity of the oxygen in oxygen carriers. In this case, selection of the oxygen carrier, which relies on the understanding of reaction mechanism of methane selective oxidation in the absence of the gaseous oxygen, is considered as one of the most essential components of the CLSOM process.

For the CLC process, it is proposed that the oxygen carriers must own the following properties in chemistry [7,8]: (i) sufficient oxygen storage and transport capacity; (ii) high reactivity in both reduction and oxidation cycles; (iii) ability to completely combust a fuel; (iv) ability of resistant to agglomeration and carbon deposition. This list also applies to the CLSOM oxygen carriers except the third one (iii), which should be changed to "ability to selectively oxidize a fuel."

Most of previous technical literatures on CLSOM focused on development of suitable oxygen carrier materials for methane selective oxidation. Ceria-based materials and perovskite-type oxides were paid the most attention due to their high lattice oxygen activity, excellent redox properties, and good thermal stability.

CeO_2-Based Oxygen Carriers

The selective oxidation of methane to CO and H_2 (syngas) by gas-solid reaction was firstly achieved over CeO_2 oxygen carrier [11, 12]. The reaction between methane and CeO_2 may occur in four equations:

$$8CeO_2 + CH_4 \longrightarrow 4Ce_2O_3 + CO_2 + H_2O, \quad (1)$$

$$2CeO_2 + CH_4 \longrightarrow Ce_2O_3 + CO + 2H_2, \quad (2)$$

$$CeO_2 + CH_4 \longrightarrow CeO_{1.83} + CO_2 + H_2O, \quad (3)$$

$$CeO_2 + CH_4 \longrightarrow CeO_{1.83} + CO + H_2. \quad (4)$$

The thermodynamic considerations of the reactions in (1)–(4) were shown in Figure 2. It is clear that the complete oxidation of methane to CO_2 and H_2O by CeO_2 (reaction (1)) is thermodynamically unfeasible under 1000°C, and the syngas generation through selective oxidation of methane by CeO_2 is favorable with the reaction temperatures ≥ 700°C. The experimental results supported the thermodynamic analysis [12]. It shows that syngas with H_2/CO ratio of 2.0 was indeed produced via the gas-solid reaction between methane and CeO_2 at 700°C, and the reduction degree of CeO_2 reached 21% with platinum as a catalyst, suggesting that almost all the CeO_2 was reduced to Ce_2O_3. This indicates that the oxidation of methane over CeO_2 may occur follow (2) in the presence of platinum.

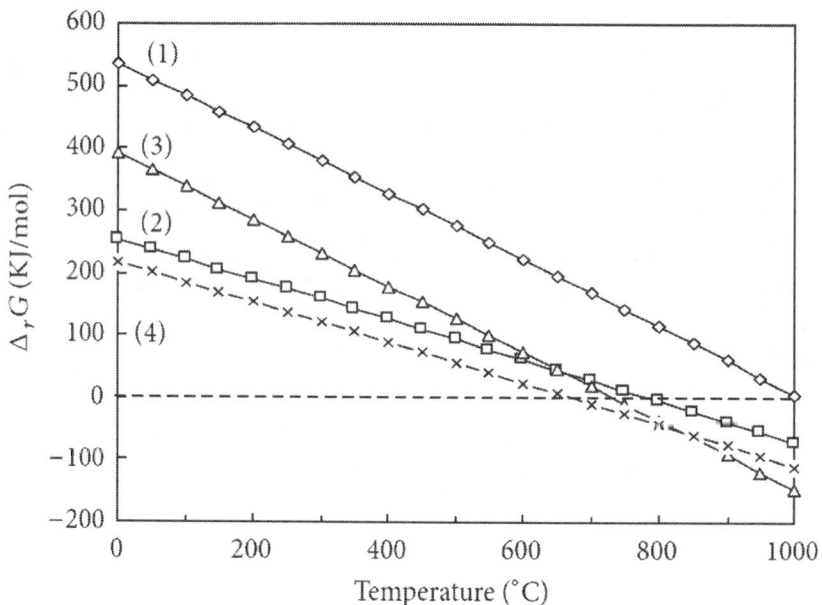

Figure 2: Thermodynamic calculations for the possible reactions between methane and CeO_2 [14].

Fathi et al. [13] also investigated the reaction between methane and CeO_2 with -Al_2O_3 as a support and Pt or Rh as a promoter. They observed that the selectivity to syngas depends on the reduction degree of CeO_2. Numbers of CO_2 and H_2O were produce in the early stage of the reaction, and then the syngas selectivity increased quickly with the reduction degree of cerium oxide. Pt or Rh promoters could lower the temperature necessary to reduce the cerium oxide but also result in the formation of carbon deposition. Pantu et al. [6] found that the surface area of Pt/CeO_2 sample affects the formation rate of syngas: methane conversion slightly increased, and syngas selectivity slightly decreased with increasing surface area. This indicates that ether high or low surface area of oxygen carrier will reduce the yield of syngas. They also observed that there was no significant effect of Pt loading on the activity of CeO_2 for methane oxidation, and the differences in metal dispersion on CeO_2 are not substantial.

Wei et al. [14] investigated the effects of CeO_2 loading on the reactivity of CeO_2/-Al_2O_3 oxygen carrier for methane selective oxidation in the absence of platinum catalyst. The results showed that higher CeO_2 loading will seriously decrease the selectivity of syngas. The other major innovation in this paper is the use of molten salt system as thermal carrier, which can avoid the agglomeration of circulating particles and improve the thermal efficiency of the whole reaction system.

It is generally accepted that the addition of Zr^{4+} could enhance the oxygen storage capacity by increasing the oxygen vacancies of ceria. Otsuka et al. [15] tested the reactivity of $Ce_{1-x}Zr_xO_2$ for the direct conversion of methane to syngas by gas-solid reactions. The formation rates of H_2 and CO were increased, and the activation energy was remarkably decreased due to the incorporation of ZrO_2 into CeO_2. The conversion of CH_4 to H_2 and CO could be achieved at a temperature as low as 500°C by using $Ce_{0.8}Zr_{0.2}O_2$ in the presence of Pt, which is 200°C lower than CeO_2 sample. Pantu et al. [6] observed that addition of ZrO_2 to CeO_2 significantly increases the methane oxidation rate and the reducibility of the CeO_2 but decreases the selectivity to H_2 and CO. Wei et al. [16] also reported a similar observation on using $Ce_{1-x}Zr_xO_2$ as oxygen carrier, but they found that the ZrO_2-rich materials own better activity and stability.

Kang and Eyring [17, 18] investigated the activity of the Ce-Zr-Tb-O system for methane oxidation and found that the oxygen transfer capacity and the oxygen storage capacity are equally important for syngas generation. The reactivity of ceria-zirconia oxides doped by Pr, Gd, or La for methane conversion was also investigated by CH_4-TPR technology and pulse reduction experiments, and it is proved that Pr-doped sample showed good activity for syngas generation [19]. The reaction between methane and Ce–Zr–Pr–O oxygen carrier with Pt as catalyst at high temperatures is controlled by the lattice oxygen diffusion, while the reactivity of weak bound surface oxygen determine the activity of the mixed oxides at the lowest temperature (~550°C).

Sadykov et al. [20] designed incorporating Sm^{3+} and Bi^{3+} cations into the ceria lattice to enhance the oxygen mobility while increasing the rate of methane dissociation by supporting Pt, and the results were also compared with the Pt/Ce-Zr-La-O mixed oxides. It showed that only the Ce-Sm-based oxide system is promising for methane selective oxidation by gas-solid reaction due to a high mobility and reactivity of the lattice oxygen, good selectivity for syngas generation, and high stability in redox cycles. The selective conversion of methane into syngas by lattice oxygen depends not only on the route of its primary activation (i.e., on supported Pt clusters) but on the features of activated fragments transformation on the support surface as well, provided the lattice oxygen mobility that is comparable.

Several reports showed that the oxidation activity and redox property of the ceria can be strongly enhanced by the addition of Fe^{3+} due to the formation of surface structural defects and Ce-Fe solid solution [21–25]. In addition, the modified iron oxides can also produce CO and H_2 through reduction with methane in an appropriate condition [26, 27], and that the iron species can strongly enhance the adsorption of methane [28]. Fe_2O_3 is possibly the most common and one of the cheapest metal oxides available in nature, and Fe^{3+} are very suitable as an dopant to improve the performance of ceria [29, 30]. Combination of CeO_2 and Fe_2O_3 gives people very high expectation to obtain attractive oxygen carriers for methane selective oxidation. Given the above, the investigation on the possibility of using CeO_2-Fe_2O_3 composite as oxygen carrier for methane selective oxidation attracted much attention [31–41].

It was reported that the CeO_2-Fe_2O_3 mixed oxides own good activity, selectivity, and stability for syngas generation through gas-solid reactions, as shown in Figure 3, and the interaction between exposed Fe_2O_3 and Ce-Fe solid solution in the oxygen carrier plays an important role on the syngas generation [36]. In addition, the dispersion of surface Fe_2O_3 and the formation of the Ce-Fe solid solution were enhanced by the redox treatment, which made the oxygen carrier very stable in the successive generation of syngas [36]. The selectivity of Ce-Fe mixed oxides for syngas production is strongly affected by the specific surface area of oxygen carriers, and high surface area would result in abundant surface adsorbed oxygen, favoring the complete oxidation of methane to carbon dioxide and water [42].

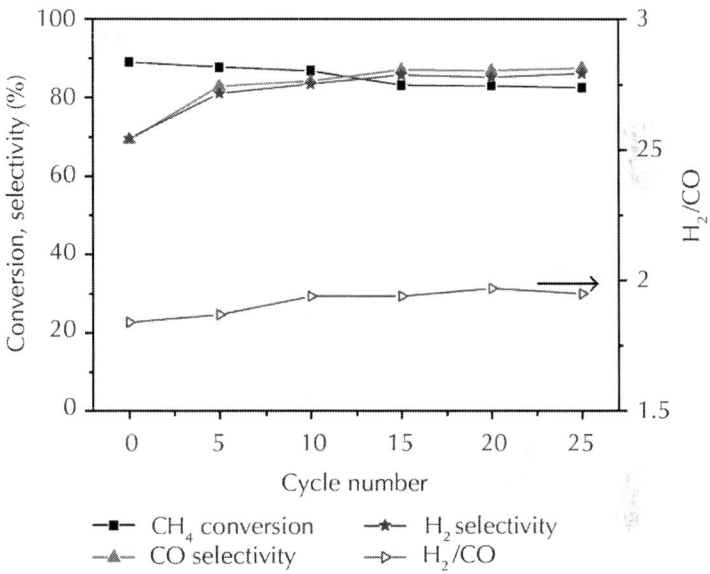

Figure 3: The effect of redox cycle number on the selective oxidation of methane using $Ce_{0.7}Fe_{0.3}O_2$ oxygen carrier at 850°C [36].

For the reaction process between methane and Ce-Fe mixed oxides [36], methane was found to adsorb and activate on the reduced iron and cerium sites, and the subsequent oxidation of activated methane relied on the lattice oxygen mobility of the oxygen carrier. The dispersion of surface iron species and the consistence of oxygen vacancy in Ce-Fe mixed oxides in turn markedly affect the formation rate of syngas, and the strong interactions between dispersed Fe species and Ce-Fe solid solution have a distinct positive effect on the catalytic activity for methane selective oxidation.

Comparison of Ce-Zr and Ce-Fe mixed oxides demonstrated that the two samples showed similar activity for methane oxidation, but the Ce-Fe sample revealed higher selectivity of syngas, as shown in Figure 4 [33]. Addition of ZrO_2 into CeO_2-Fe_2O_3 system could enhance the interaction between iron and cerium oxides via increasing the oxygen vacancy concentration and improving the dispersion of free Fe_2O_3, which improved the activity of Ce-Fe mixed oxides for methane selective oxidation. However, heavy loading of ZrO_2 would lead to a phase segregation of CeO_2 and Fe_2O_3 from the Ce-Fe solid solution, resulting in a decrease in syngas selectivity [34].

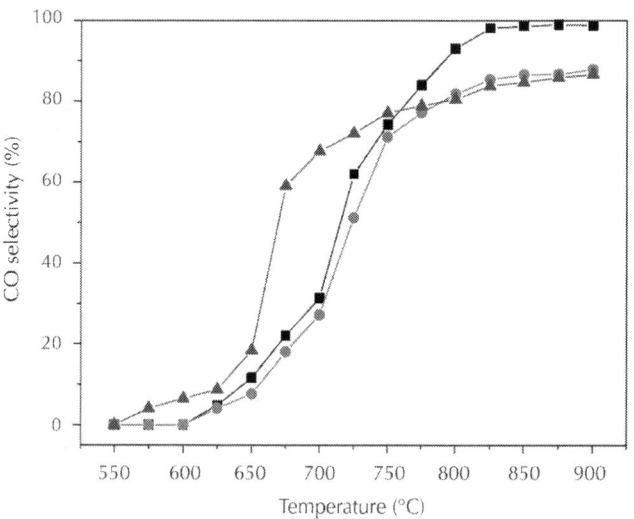

Figure 4: CO and H_2 selectivity as a function of reaction temperature over CeO_2-Fe_2O_3, CeO_2-ZrO_2, and ZrO_2-Fe_2O_3 oxygen carriers [33].

The effect of supports (Al_2O_3, SiO_2, and MgO) on the activity and selectivity of Ce-Fe-Zr mixed oxides for methane selective oxidation was also investigated [40]. Al_2O_3 support could result in the complete oxidation of methane, and SiO_2 obvious reduced the reactivity of Ce-Fe-Zr mixed oxides. On the other hand, MgO support strongly enhanced the activity and selectivity of Ce-Fe-Zr oxygen carriers for syngas generation.

Ce-Cu-O, Ce-Mn-O, and Ce-Nb-O and Ce-Ni-O were also considered as an oxygen carrier for methane oxidation [38, 43–45]. Compared with the Ce-Fe mixed oxides, Ce-Cu-O and Ce-Mn-O oxygen carriers are more favorable to completely oxidize methane [38], and the Ce-Ni-O would result in the decomposition of methane when the Ni loading is too high [44]. For Ce-Nb-O system, the further oxidation of hydrogen to water was observed, and the CO and H_2O were the main production [45]. For all ceria-based oxygen carriers, the reoxidation process by air is very easy to accomplish.

Perovskite Oxygen Carriers

Perovskite oxides with an ABO_3-type crystal structure usually exhibit excellent redox properties, high oxygen mobility, and thermal stability, which can be used in many reactions related to a redox process, such as catalytic purification of automotive exhaust and solid oxide fuel cell (SOFC) [46–50]. As a famous perovskite oxides, $LaFeO_3$ was firstly chosen to selectively oxidize methane by Dai et al. [51, 52], and its performance was compared with $NdFeO_3$ and $EuFeO_3$. The oxygen storage and transport ability of $AFeO_3$ (A = La, Nd, and Eu) is related to its Fe–O bond distance and shorter distance given lower activity of oxygen. The reaction rate between methane $AFeO_3$ strongly depends on the reaction temperature, and high-temperature (>800°C) is necessary for obtaining a high syngas yield. The $LaFeO_3$ oxide exhibits the best performance among these tested $AFeO_3$ oxides (A = La, Nd, and Eu) for syngas production, and it also maintains high catalytic activity and structural stability in the redox experiment between methane and air at 900°C. It was also observed that the reduction of $LaFeO_3$ by methane was performed through a reduction of Fe^{3+} to Fe^{2+}, and further reduction is very difficult [53]. They also investigated the redox property of $LaFeO_3$ for successive generation of syngas in a circulating-fluidized bed (CFB) reactor [54]. It showed that methane could be oxidized to syngas by lattice oxygen with high selectivity, and the depleted oxygen species could be regenerated in a CFB condition. The methane conversion remains at 60%–70% with the CO selectivity of ca. 96% during the 30 redox cycles. However, this paper did not involve the mechanical performance of the oxygen carriers in the redox process, and it is proposed that the attrition resistance for CFB process should be paid much attention.

Li et al. [55, 56] added Sr into the $LaFeO_3$ system to partially substitute the sites of La and investigated that performance of the $La_{0.8}Sr_{0.2}FeO_3$ oxide for methane selective oxidation. They proposed that there are two kinds of oxygen species on the oxide: (i) the active oxygen species (weakly bound oxygen species) which are responsible for complete oxidation of methane and (ii) the weaker oxygen species (strongly bound oxygen species) which are responsible for partial oxidation of methane to syngas. This is similar with the observation by Greish et al. [57]. Methane reacts firstly with the active oxygen species to form CO_2 and H_2O, and then the weaker oxygen species can oxidize methane to CO and H_2 with high selectivity.

On the other hand, substituting La for Sr was found to increase the oxygen capacity of these materials but reduce the selectivity to syngas and the reactivity with CH_4 [58]. Addition of Cr, Ni, and Cu into the La-Sr-Fe-O perovskite system to partially replace the Fe sites could improve the reactivity for methane conversion [59], while incorporation of Co ions into La-Sr-Fe-O mixed oxides could enhance the activity of this material for methane combustion [58] and reduce the stability under redox testing [60]. The $La_{0.7}Sr_{0.3}Cr_{0.1}Fe_{0.9}O_3$ with physically mixed NiO as a catalyst showed good activity and stability in the redox testing [59]. The improvement on the syngas production and stability of material was also observed over the NiO $La_{1-x}Sr_xFeO_3$/system due to the presence of exposed NiO particles, but the presence of NiO also improved the catalytic activity for methane decomposition, resulting in the formation of carbon deposition [61].

$La_{1-x}Sr_xFeO_3$ (M = Mn, Ni) and $LaMnO_{3-\delta}F$ perovskite oxides were also investigated as oxygen carriers for methane oxidation [62]. It is proposed that the reactivity and selectivity of lattice oxygen depend on (i) B-site element, (ii) degree of substitution of La with Sr, and (iii) fluorination of the perovskite oxide. $La_{1-x}Sr_xFeO_3$ The with relatively low degree of Sr-substitution and the fluorinated $LaMnO_{3-\delta}F$ are suitable oxygen carriers for syngas generation. The high substitution degree of La by Sr increases the reactivity of lattice oxygen but decreases the selectivity to syngas. Evdou et al. [61] observed that the reduction degree $La_{1-x}Sr_xFeO_3$ of oxides by methane relies on the Sr content and the reaction temperature.

$BaTi_{1-x}In_xO_3$ Perovskite oxides with nickel as a catalyst were also investigated for methane oxidation in the absence of gas phase oxygen

[63]. Based on the temperature-programmed surface reaction of methane (TPSR-CH_4) and pulses reaction results, they found that the reducibility of B cation (ABO_3) and the anionic conductivity of the material strongly influence the activity and selectivity $BaTi_{1-x}In_xO_3$ of oxygen carriers. It is also observed that $Ni/BaTi_{0.3}In_{0.7}O_3$ oxygen carrier was more stable than $Ni/BaTiO_3$ due to the existence of Ni-In alloys which is relatively inert for catalyzing the cracking reaction.

Other Oxygen Carriers

Fe_2O_3 as oxygen carrier was proved to own the ability for methane combustion [7, 8], but addition of other suitable oxides can modify the selectivity of its lattice oxygen for selective oxidation of methane to syngas. Fe_2O_3-Rh_2O_3/Y_2O_3 and Fe_2O_3-Cr_2O_3-MgO oxides were found to be active to produce syngas with a moderate methane conversion and selectivity [64]. It is also proved that combining CuO and Fe_2O_3 to form a Cu-ferrite could obtain a suitable oxygen carrier for methane selective oxidation. Cha et al. [65, 66] investigated the reactivity of /Ce-ZrO_2 () for methane selective oxidation, and the results showed that the Cu-ferrite suppressed carbon deposition and promoted the reactivity with methane to produce syngas. On the other hand, since the lattice oxygen from Fe_2O_3 to Fe_3O_4 can completely oxidize methane to CO_2 and H_2O, the redox of FeO/Fe_3O_4 was proposed to convert methane to syngas by a chemical looping step [67].

NiO-based materials were also used for the methane selective oxidation, but significant amount of CO_2 and H_2O was observed in the products over NiO, NiO/ -Al_2O_3, NiO/ -Al_2O_3, and NiO/Mg-ZrO_2 oxygen carriers [68–73]. During the reaction between methane and NiO-based oxides, the syngas yield depends on the oxidation degree of the oxygen carriers: highly oxidized oxide particles resulted in the formation of CO_2 and H_2O, while reduced particles could produce CO and H_2 [70, 74]. Addition of Cr_2O_3 into NiO-MgO system could change reactivity of the lattice oxygen in the materials, and the fully oxidized NiO-Cr_2O_3-MgO produced H_2 and CO with high selectivity during the reaction with CH_4 [64, 75]. The appearance of $NiAl_2O_4$ also could reduce the activity of oxygen in the material and promote the formation of H_2 and CO [76].

Based on the previous discussions, CeO_2-based oxygen carrier could convert methane into syngas at relatively low temperatures (ca. 700°C) in the presence of Pt promoter, but the redox stability of the oxygen carriers needs to be improved. The perovskite-type oxygen carriers own high selectivity and redox stability for syngas generation, but they are only active at high temperatures (ca. 850°C). For the Fe_2O_3- and NiO-based oxygen carriers, a large number of CO_2 and H_2O were produced during the gas-solid reaction between oxygen carrier and methane. Although addition of suitable promoters could improve the selectivity of oxygen carriers for syngas generation, but it also reduced the reactivity for methane conversion. Among the different oxygen carriers, perovskite-type oxygen carriers are more competitive for the CLSOM process, if the activity could be enhanced by the structure modifications. On the other hand, the use of the various combinations of catalysts (e.g., combinations of perovskite-type or CeO_2-ZrO_2 oxygen carriers with Ni or Fe species) may also achieve the greater efficiency for syngas generation.

CONCLUSIONS

Chemical-looping selective oxidation of methane (CLSOM) is a promising, energy-efficient, and low-cost route for syngas generation. However, at the present time, this technology is not fully established for large-scale implementation, and valuable researches need to be developed to address the important issues of this technology. Nowadays, numbers of works were performed on this technology, and most of previous technical literatures had been focused on the development of suitable oxygen carrier materials. After reviewing such references, it is found that a suitable oxygen carrier should own abundant active sites for methane activation, high oxygen storage capacity, and good oxygen mobility. This finding gives useful references for the further developing highly efficient oxygen carriers.

Due to the two-step redox process, the chemical engineering of whole process is actually a key factor for success in practical application, and the specific selected reactor design is very critical. The mechanical performance of the oxygen carriers should be paid much attention when a fluidized bed reactor is used. In addition, since the reaction between methane and oxygen carriers is endothermic, while the reoxidation of reduced oxygen carriers is an exothermic reaction, the energy efficiency of the whole process strongly depends on the transfer of the heat from the exothermic reaction to the endothermic reaction. This issue is also very important for the practical application of this technology.

ACKNOWLEDGMENTS

This paper was supported by the National Nature Science Foundation of China (Project nos. 51004060 and 51174105), National Excellent Doctoral Dissertation Development Foundation of Kunming University of Science and Technology, Natural Science Foundation of Yunnan Province (no. 2010ZC018), and a school-enterprise cooperation project from Jinchuan Corporation (no. Jinchuan 201115).

REFERENCES

1. J. R. Rostrup-Nielsen, "Fuels and energy for the future: the role of catalysis," Catalysis Reviews, vol. 46, no. 3-4, pp. 247–270, 2004.
2. A. Holmen, "Direct conversion of methane to fuels and chemicals," Catalysis Today, vol. 142, no. 1-2, pp. 2–8, 2009.
3. B. C. Enger, R. Lødeng, and A. Holmen, "A review of catalytic partial oxidation of methane to synthesis gas with emphasis on reaction mechanisms over transition metal catalysts," Applied Catalysis A, vol. 346, no. 1-2, pp. 1–27, 2008.
4. P. M. Torniainen, X. Chu, and L. D. Schmidt, "Comparison of monolith-supported metals for the direct oxidation of methane to syngas," Journal of Catalysis, vol. 146, no. 1, pp. 1–10, 1994.
5. L. Bobrova, N. Vernikovskaya, and V. Sadykov, "Conversion of hydrocarbon fuels to syngas in a short contact time catalytic reactor," Catalysis Today, vol. 144, no. 3-4, pp. 185–200, 2009.
6. P. Pantu, K. Kim, and G. R. Gavalas, "Methane partial oxidation on Pt/CeO2-ZrO2 in the absence of gaseous oxygen," Applied Catalysis A, vol. 193, no. 1-2, pp. 203–214, 2000.

7. J. Adanez, A. Abad, F. Garcia-Labiano, P. Gayan, and L. F. de Diego, "Progress in chemical-looping combustion and reforming technologies," Progress in Energy and Combustion Science, vol. 38, no. 2, pp. 215–282, 2012.
8. M. M. Hossain and H. I. de Lasa, "Chemical-looping combustion (CLC) for inherent CO2 separations-a review," Chemical Engineering Science, vol. 63, no. 18, pp. 4433–4451, 2008.
9. F. vanlooij, J. C. van Giezen, E. R. Stobbe, and J. W. Geus, "Mechanism of the partial oxidation of methane to synthesis gas on a silica-supported nickel catalyst," Catalysis Today, vol. 21, no. 2-3, pp. 495–503, 1994.
10. E. R. Stobbe, B. A. De Boer, and J. W. Geus, "The reduction and oxidation behaviour of manganese oxides," Catalysis Today, vol. 47, no. 1–4, pp. 161–167, 1999.
11. K. Otsuka, T. Ushiyama, and I. Yamanaka, "Partial oxidation of methane using the redox of cerium oxide," Chemistry Letters, pp. 1517–1520, 1993.
12. K. Otsuka, Y. Wang, E. Sunada, and I. Yamanaka, "Direct partial oxidation of methane to synthesis gas by cerium oxide," Journal of Catalysis, vol. 175, no. 2, pp. 152–160, 1998.
13. M. Fathi, E. Bjorgum, T. Viig, and O. A. Rokstad, "Partial oxidation of methane to synthesis gas: elimination of gas phase oxygen," Catalysis Today, vol. 63, no. 2–4, pp. 489–497, 2000.
14. Y. Wei, H. Wang, F. He, X. Ao, and C. Zhang, "CeO2 as the oxygen carrier for partial oxidation of methane to synthesis gas in molten salts: thermodynamic analysis and experimental investigation," Journal of Natural Gas Chemistry, vol. 16, no. 1, pp. 6–11, 2007.
15. K. Otsuka, Y. Wang, and M. Nakamura, "Direct conversion of methane to synthesis gas through gas-solid reaction using CeO2-ZrO2 solid solution at moderate temperature," Applied Catalysis A, vol. 183, no. 2, pp. 317–324, 1999.
16. Y. G. Wei, H. Wang, K. Z. Li, M. C. Liu, and X. Q. Ao, "Preparation and performance of Ce/Zr mixed oxides for direct conversion of methane to syngas," Journal of Rare Earths, vol. 25, pp. 110–114, 2007.
17. Z. C. Kang and L. Eyring, "Hydrogen production from methane and water by lattice oxygen transfer with $Ce0.70Zr0.25Tb0.05O_{2-x}$,"

Journal of Alloys and Compounds, vol. 323-324, pp. 97–101, 2001.

18. Z. C. Kang and L. Eyring, "Lattice oxygen transfer in fluoritetype oxides containing Ce, Pr, and/or Tb," Journal of Solid State Chemistry, vol. 155, no. 1, pp. 129–137, 2000.

19. V. A. Sadykov, N. N. Sazonova, A. S. Bobin et al., "Partial oxidation of methane on Pt-supported lanthanide doped ceriazirconia oxides: effect of the surface/lattice oxygen mobility on catalytic performance," Catalysis Today, vol. 169, no. 1, pp. 125–137, 2011.

20. V. A. Sadykov, T. G. Kuznetsova, G. M. Alikina et al., "Ceriabased fluorite-like oxide solid solutions as catalysts of methane Journal of Chemistry 7 selective oxidation into syngas by the lattice oxygen: synthesis, characterization and performance," Catalysis Today, vol. 93–95, pp. 45–53, 2004.

21. C. Liu, L. Luo, and X. Lu, "Preparation of mesoporous $Ce_{1-x}Fe_xO_2$ mixed oxides and their catalytic properties in methane combustion," Kinetics and Catalysis, vol. 49, no. 5, pp. 676–681, 2008.

22. J. Y. Luo, M. Meng, J. S. Yao et al., "One-step synthesis of nanostructured Pd-doped mixed oxides MO_x-CeO_2 (M = Mn, Fe, Co, Ni, Cu) for efficient CO and C_3H_8 total oxidation," Applied Catalysis B, vol. 87, no. 1-2, pp. 92–103, 2009.

23. H. Lv, H. Y. Tu, B. Y. Zhao, Y. J. Wu, and K. A. Hu, "Synthesis and electrochemical behavior of $Ce_{1-x}Fe_xO_{2-\delta}$ as a possible SOFC anode materials," Solid State Ionics, vol. 177, no. 39-40, pp. 3467–3472, 2007.

24. H. Lv, D. J. Yang, X. M. Pan et al., "Performance of Ce/Fe oxide anodes for SOFC operating on methane fuel," Materials Research Bulletin, vol. 44, no. 6, pp. 1244–1248, 2009.

25. C. Liang, Z. Ma, H. Lin et al., "Template preparation of nanoscale $Ce_xFe_{1-x}O_2$ solid solutions and their catalytic properties for ethanol steam reforming," Journal of Materials Chemistry, vol. 19, no. 10, pp. 1417–1424, 2009.

26. S. Takenaka, M. Serizawa, and K. Otsuka, "Formation of filamentous carbons over supported Fe catalysts through methane decomposition," Journal of Catalysis, vol. 222, no. 2, pp. 520–531, 2004.

27. O. Nakayama, N. O. Ikenaga, T. Miyake, E. Yagasaki, and T. Suzuki, "Partial oxidation of CH4 with air to produce pure hydrogen and syngas," Catalysis Today, vol. 138, no. 3-4, pp. 141–146, 2008.
28. S. Fukuda, T. Hino, and T. Yamashina, "Desorption processes of hydrogen and methane from clean and metal-deposited graphite irradiated by hydrogen ions," Journal of Nuclear Materials, vol. 162–164, no. C, pp. 997–1003, 1989.
29. G. Li, R. L. Smith, and H. Inomata, "Synthesis of nanoscale Ce1−xFexO2 solid solutions via a low-temperature approach," Journal of the American Chemical Society, vol. 123, no. 44, pp. 11091–11092, 2001.
30. F. J. Perez-Alonso, M. L. Granados, M. Ojeda et al., "Chemical structures of coprecipitated Fe-Ce mixed oxides," Chemistry of Materials, vol. 17, no. 9, pp. 2329–2339, 2005.
31. K. Li, H. Wang, Y. Wei, and M. Liu, "Catalytic performance of cerium iron complex oxides for partial oxidation of methane to synthesis gas," Journal of Rare Earths, vol. 26, no. 5, pp. 705–710, 2008.
32. K. Li, H. Wang, Y. Wei, and M. Liu, "Preparation and characterization of Ce1−xFexO2 complex oxides and its catalytic activity for methane selective oxidation," Journal of Rare Earths, vol. 26, no. 2, pp. 245–249, 2008.
33. K. Li, H. Wang, Y. Wei, and D. Yan, "Direct conversion of methane to synthesis gas using lattice oxygen of CeO2-Fe2O3 complex oxides," Chemical Engineering Journal, vol. 156, no. 3, pp. 512–518, 2010.
34. K. Z. Li, H. Wang, Y. G. Wei, and D. X. Yan, "Partial oxidation of methane to syngas with air by lattice oxygen transfer over ZrO2-modified Ce-Fe mixed oxides," Chemical Engineering Journal, vol. 173, pp. 574–582, 2011.
35. Y. Wei, H. Wang, and K. Li, "Ce-Fe-O mixed oxide as oxygen carrier for the direct partial oxidation of methane to syngas," Journal of Rare Earths, vol. 28, no. 4, pp. 560–565, 2010.
36. K. Li, H. Wang, Y. Wei, and D. Yan, "Syngas production from methane and air via a redox process using Ce-Fe mixed oxides as oxygen carriers," Applied Catalysis B, vol. 97, no. 3-4, pp. 361–372, 2010.

37. K. Li, H. Wang, Y. Wei, X. Ao, and M. Liu, "Partial oxidation of methane to synthesis gas using lattice oxygen," Progress in Chemistry, vol. 20, no. 9, pp. 1306–1314, 2008.
38. F. He, Y. Wei, H. Li, and H. Wang, "Synthesis gas generation by Chemical-looping reforming using Ce-based oxygen carriers modified with Fe, Cu, and Mn oxides," Energy and Fuels, vol. 23, no. 4, pp. 2095–2102, 2009.
39. X. Zhu, H. Wang, Y. Wei, K. Li, and X. Cheng, "Hydrogen and syngas production from two-step steam reforming of methane over CeO_2-Fe_2O_3 oxygen carrier," Journal of Rare Earths, vol. 28, no. 6, pp. 907–913, 2010.
40. X. Cheng, H. Wang, Y. Wei, K. Li, and X. Zhu, "Preparation and characterization of Ce-Fe-Zr-O(x)/MgO complex oxides for selective oxidation of methane to synthesize gas," Journal of Rare Earths, vol. 28, no. 1, pp. 316–321, 2010.
41. H. Kaneko, H. Ishihara, S. Taku, Y. Naganuma, N. Hasegawa, and Y. Tamaura, "Cerium ion redox system in $CeO_2-xFe_2O_3$ solid solution at high temperatures (1,273–1,673 K) in the twostep water-splitting reaction for solar H_2 generation," Journal of Materials Science, vol. 43, no. 9, pp. 3153–3161, 2008.
42. K. Li, H. Wang, Y. Wei, and D. Yan, "Transformation of methane into synthesis gas using the redox property of Ce-Fe mixed oxides: effect of calcination temperature," International Journal of Hydrogen Energy, vol. 36, no. 5, pp. 3471–3482, 2011.
43. K. Z. Li, H. Wang, Y. G. Wei, and M. C. Liu, "Partial oxidation of methane to syngas using lattice oxygen from ceria-based complex oxides oxygen carriers," Journal of Fuel Chemistry and Technology, vol. 36, no. 1, pp. 83–88, 2008.
44. Y. Wei, H. Wang, K. Li, X. Zhu, and Y. Du, "Preparation and characterization of $Ce_{1-x}NiO_2$ as oxygen carrier for selective oxidation methane to syngas in absence of gaseous oxygen," Journal of Rare Earths, vol. 28, no. 1, pp. 357–361, 2010.
45. A. A. Yaremchenko, V. V. Kharton, S. A. Veniaminov, V. D. Belyaev, V. A. Sobyanin, and F. M. B. Marques, "Methane oxidation by lattice oxygen of $CeNbO_{4+\delta}$," Catalysis Communications, vol. 8, no. 3, pp. 335–339, 2007.

46. U. Balachandran, J. T. Dusek, R. L. Mieville et al., "Dense ceramic membranes for partial oxidation of methane to syngas," Applied Catalysis A, vol. 133, no. 1, pp. 19–29, 1995.
47. V. R. Choudhary, S. Banerjee, and B. S. Uphade, "Activation by hydrothermal treatment of low surface area ABO3-type perovskite oxide catalysts," Applied Catalysis A, vol. 197, no. 2, pp. L183–L186, 2000.
48. N. E. Trofimenko and H. Ullmann, "Oxygen stoichiometry and mixed ionic-electronic conductivity of $Sr_{1-a}Ce_aFe_{1-b}Co_bO_{3-x}$ perovskite-type oxides," Journal of the European Ceramic Society, vol. 20, no. 9, pp. 1241–1250, 2000.
49. M. van den Bossche and S. McIntosh, "The rate and selectivity of methane oxidation over $La_{0.75}Sr_{0.25}Cr_xMn_{1-x}O_{3-\delta}$ as a function of lattice oxygen stoichiometry under solid oxide fuel cell anode conditions," Journal of Catalysis, vol. 255, no. 2, pp. 313–323, 2008.
50. A. Khanfekr, K. Arzani, A. Nemati, and M. Hosseini, "Production of perovskite catalysts on ceramic monoliths with nanoparticles for dual fuel system automobiles," International Journal of Environmental Science and Technology, vol. 6, no. 1, pp. 105–112, 2009. 8 Journal of Chemistry
51. X. P. Dai, R. J. Li, C. C. Yu, and Z. P. Hao, "Unsteady-state direct partial oxidation of methane to synthesis gas in a fixed-bed reactor using AFeO3 (A = La, Nd, Eu) perovskite-type oxides as oxygen storage," Journal of Physical Chemistry B, vol. 110, no. 45, pp. 22525–22531, 2006.
52. X. P. Dai and C. C. Yu, "Nano-perovskite-based (LaMO3) oxygen carrier for syngas generation by chemical-looping reforming of methane," Chinese Journal of Catalysis, vol. 32, pp. 1411–1417, 2011.
53. O. Mihai, D. Chen, and A. Holmen, "Catalytic consequence of oxygen of lanthanum ferrite perovskite in chemical looping reforming of methane," Industrial and Engineering Chemistry Research, vol. 50, no. 5, pp. 2613–2621, 2011.
54. X. Dai, C. Yu, R. Li, Q. Wu, and Z. Hao, "Synthesis gas production using oxygen storage materials as oxygen carrier over circulating fluidized bed," Journal of Rare Earths, vol. 26, no. 1, pp. 76–80, 2008.

55. R. Li, C. Yu, X. Dai, and S. Shen, "Selective oxidation of methane to synthesis gas using lattice oxygen from perovskite La0.8Sr0.2FeO3 catalyst," Chinese Journal of Catalysis, vol. 23, no. 6, pp. 549–554, 2002.
56. R. J. Li, C. C. Yu, W. J. Ji, and S. K. Shen, "Methane oxidation to synthesis gas using lattice oxygen in La1−xSrxFeO3 perovskite oxides instead of molecular oxygen," Studies in Surface Science and Catalysis, vol. 147, pp. 199–204, 2004.
57. A. A. Greish, L. M. Glukhov, E. D. Finashina et al., "Oxidative coupling of methane in the redox cyclic mode over the catalysts on the basis of CeO2 and La2O2," Mendeleev Communications, vol. 20, no. 1, pp. 28–30, 2010.
58. M. Ryden, A. Lyngfelt, T. Mattisson, D. Chen, A. Holmen, and ´E. Bjørgum, "Novel oxygen-carrier materials for chemicallooping combustion and chemical-looping reforming; LaxSr1−xFeyCo1−yO3−δ perovskites and mixed-metal oxides of NiO, Fe2O3 and Mn3O4," International Journal of Greenhouse Gas Control, vol. 2, no. 1, pp. 21–36, 2008.
59. L. Nalbandian, A. Evdou, and V. Zaspalis, "La1−xSrxMyFe1−yO3−δ perovskites as oxygen-carrier materials for chemical-looping reforming," International Journal of Hydrogen Energy, vol. 36, no. 11, pp. 6657–6670, 2011.
60. A. Murugan, A. Thursfield, and I. S. Metcalfe, "A chemical looping process for hydrogen production using iron-containing perovskites," Energy & Environmental Science, vol. 4, pp. 4639–4649, 2011.
61. A. Evdou, V. Zaspalis, and L. Nalbandian, "La1−xSrxFeO3−δ perovskites as redox materials for application in a membrane reactor for simultaneous production of pure hydrogen and synthesis gas," Fuel, vol. 89, no. 6, pp. 1265–1273, 2010.
62. H. J. Wei, Y. Cao, W. J. Ji, and C. T. Au, "Lattice oxygen of La1−xSrxMO3 (M = Mn, Ni) and LaMnO3−$\alpha F\beta$ perovskite oxides for the partial oxidation of methane to synthesis gas," Catalysis Communications, vol. 9, no. 15, pp. 2509–2514, 2008.
63. V. Garc´ıa, M. T. Caldes, O. Joubert, E. Gautron, F. Mondragon, ´and A. Moreno, "Methane oxidation by lattice oxygen of Ni/BaTi1−xInxO3−δ catalysts," Catalysis Today, vol. 157, no. 1–4, pp. 177–182, 2010.

64. T. Suzuki, O. Nakayama, and N. Okamoto, "Partial oxidation of methane to nitrogen free synthesis gas using air as oxidant," Catalysis Surveys From Asia, vol. 16, pp. 75–90, 2012.
65. K. S. Cha, H. S. Kim, B. K. Yoo et al., "Reaction characteristics of two-step methane reforming over a Cu-ferrite/Ce-ZrO2 medium," International Journal of Hydrogen Energy, vol. 34, no. 4, pp. 1801–1808, 2009.
66. K. S. Cha, B. K. Yoo, H. S. Kim et al., "A study on improving reactivity of Cu-ferrite/ZrO2 medium for syngas and hydrogen production from two-step thermochemical methane reforming," International Journal of Energy Research, vol. 34, no. 5, pp. 422–430, 2010.
67. K. S. Go, S. R. Son, S. D. Kim, K. S. Kang, and C. S. Park, "Hydrogen production from two-step steam methane reforming in a fluidized bed reactor," International Journal of Hydrogen Energy, vol. 34, no. 3, pp. 1301–1309, 2009.
68. L. F. de Diego, M. Ortiz, F. Garcı́a-Labiano, J. Adanez, A. Abad, and P. Gayan, "Synthesis gas generation by chemical-looping reforming using a Ni based oxygen carrier," in Proceedings of the 9th International Conference on Greenhouse Gas Control Technologies (GHGT-9), vol. 1, pp. 3–10, Washington DC, USA, November 2008.
69. L. F. de Diego, M. Ortiz, F. Garcı́a-Labiano, J. Adanez, A. Abad, and P. Gayan, "Hydrogen production by chemical-looping reforming in a circulating fluidized bed reactor using Ni-based oxygen carriers," Journal of Power Sources, vol. 192, no. 1, pp. 27–34, 2009.
70. M. Ryden, M. Johansson, A. Lyngfelt, and T. Mattisson, "NiO supported on Mg-ZrO2 as oxygen carrier for chemical-looping combustion and chemical-looping reforming," Energy and Environmental Science, vol. 2, no. 9, pp. 970–981, 2009.
71. M. Ortiz, L. F. de Diego, A. Abad, F. Garcı́a-Labiano, P. Gayan, and J. Adanez, "Hydrogen production by auto-thermal chemical-looping reforming in a pressurized fluidized bed reactor using Ni-based oxygen carriers," International Journal of Hydrogen Energy, vol. 35, no. 1, pp. 151–160, 2010.
72. T. Proll, J. Bolhär-Nordenkampf, P. Kolbitsch, and H. Hofbauer, "Syngas and a separate nitrogen/argon stream via chemical

looping reforming—a 140 kW pilot plant study," Fuel, vol. 89, no. 6, pp. 1249–1256, 2010.

73. M. Ryden and P. Ramos, "H2 production with CO2 capture by sorption enhanced chemical-looping reforming using NiO as oxygen carrier and CaO as CO2 sorbent," Fuel Processing Technology, vol. 96, pp. 27–36, 2012.

74. M. Ortiz, L. F. de Diego, A. Abad, F. Garcia-Labiano, P. Gayan, and J. Adanez, "Catalytic activity of Ni-based oxygen-carriers for steam methane reforming in chemical-looping processes," Energy & Fuels, vol. 26, no. 2, pp. 791–800, 2012.

75. O. Nakayama, N. Ikenaga, T. Miyake, E. Yagasaki, and T. Suzuki, "Production of synthesis gas from methane using lattice oxygen of NiO-Cr2O3-MgO complex oxide," Industrial and Engineering Chemistry Research, vol. 49, no. 2, pp. 526–534, 2010.

76. C. Dueso, M. Ortiz, A. Abad et al., "Reduction and oxidation kinetics of nickel-based oxygen-carriers for chemical-looping combustion and chemical-looping reforming," Chemical Engineering Journal, vol. 188, pp. 142–154, 2012.

Chapter 4

Synthesis of ZSM-22 in Static and Dynamic System Using Seeds

Lenivaldo V. de Sousa Júnior, Antonio O. S. Silva, Bruno J. B. Silva, and Soraya L. Alencar

Chemical Engineering Department, Federal University of Alagoas, Laboratory Synthesis Catalysts (LSCat), Maceió, Brazil

ABSTRACT

ZSM-22 was synthesized using various sources of silica, organic template 1,6-diaminohexane, under hydrothermal conditions, with and without agitation during crystallization. Subsequently, the crystallized material was used as seeds to accelerate the crystallization process. Characterization of the ZSM-22 samples was performed by XRD, ATG/DTG and FT-IR. It was found that it is possible to synthesize ZSM-22 employing colloidal silica and pyrolytic silica as silicon sources only if the system is stirred during crystallization. The crystallization time for these systems was 13 hours; longer times of crystallization do not significantly increase the crystallinity of the sample. The addition of seeds significantly accelerates the crystallization of ZSM-22, reducing the crystallization time to only 7 hours, with stirring and with systems employing colloidal silica.

INTRODUCTION

ZSM-22 belongs to the group of medium-pore zeolite (mordenite family) which includes ZSM-5, ZSM-11, and ZSM-35. It is rich in five-membered rings with wavy channels surrounded by ten-membered rings. The opening of channels (free diameter) is 0.45 × 0.55 nm running through the structure in a single crystallographic direction, having no intersections of the channels. The ten members' channels of ZSM-22 are smaller than those found in ZSM-5, ZSM-11 and ZSM-35, however the determination of the structure of ZSM-22 indicated that its topology is similar to that of zeolite Theta-1 (TON structure type) is also similar to that found in the materials: Nu-10, ISI-1, and KZ-2.

ZSM-22 in acid form is used in reactions such as: selective formation of p-dialquilbenzenes, enriched in p-xylene C8 aromatics, alkylation of toluene with methanol selective for p-xylene and isomerization of 1-butene are particularly desirable in the petroleum industry. The acidic properties and selectivity of ZSM-22 have been addressed by many researchers in their studies [1] - [4].

Seed-assisted crystallization is a very useful method and has been used since the synthesis of the first zeolites in the mid-1960s, where the main function is to provide the system with the area over which the products are to be developed, avoiding the need for generation of this surface by the system itself, through primary nucleation. Its main effect is a reduction of the synthesis time, due to increased crystallization rate, resulting in materials with low impurity content [5] [6].

This paper studied the synthesis of ZSM-22, using 1,6-diaminohexane as the organic template, various sources of silica, with the SiO_2/Al_2O_3 ratio constant. Thereafter, the zeolite crystals were used as seed for further syntheses.

The challenge is to synthesize a copy of pure ZSM-22, since ZSM-5 and cristobalite impurities are difficult to avoid [7]. Different sources of silica are studied because their dissolution and reaction temperature determine the rate of crystallization of zeolite. Colloidal silica reacts rapidly to temperatures below 170°C while the pyrolytic silica reacts readily at 170°C [8].

EXPERIMENTAL

Synthesis of Zeolite ZSM-22 with Different Sources of Silica

For synthesis the following reagents were used: 1,6-diaminohexane as organic template (Sigma-Aldrich, 98%), potassium hydroxide (Sigma-Aldrich, 85%), colloidal silica (Ludox AS40, Sigma-Aldrich), pyrolytic (Fumed) silica (0.014 µ, sigma-aldrich), silica gel (Diasil 200, Diatom), cogel from the precipitation of silica of the commercial sodium silicate (Pernambuco chemical), aluminum sulfate hydrate (Merck) and distilled water.

The gel had the following molar composition: 27 $NH_2(CH_2)_6NH_2$:13.5 $K_2O:Al_2O_3$:90 SiO_2:3600 H_2O, obtained following the methodology:

- Solution A: Potassium hydroxide is dissolved in 30% of the total amount of water required for the synthesis;
- Solution B: 1, 6-diaminohexane is solubilized in 30% of the water required for the synthesis;
- Solution C: aluminum sulphate is dissolved in 30% of the water required for the synthesis;
- Solution D: silica dispersed in the remaining water of synthesis.

The procedure for preparation of the gel consists of: 1) adding solution B to solution A; 2) addition of the solution C; and 3) finally adding the solution D. At each addition step the system was agitated vigorously (with a mechanical stirrer) for 5 min at 400 rpm. The final gel obtained remained in this condition for 30 additional minutes.

For the agitated synthesis, the gel was transferred to a Teflon vessel with a volume of 700 mL. This container was placed inside a stainless steel vessel (with a volume of 1000 mL) in a Parr reactor model 4520. The reaction mixture was stirred at 400 rpm with a double helix impeller (with 6 blades on each rotor) while the system was heated from room temperature to 160°C (within two hours).

For the static synthesis, the gel was placed in a Teflon vessel with a volume of 70 mL, and placed inside coated stainless steel autoclaves. Crystallization occurs at a temperature of 160°C for periods up to 24 hours.

Synthesis of Zeolite ZSM-22 Using Seeds

The preparation method of the mixture and the synthesis conditions are the same as those described in the previous section. The seeds are added after mixing the solutions A-D, in the amount of 0.71wt% of the synthesis gel, and stirred for 15 minutes. Thus, the molar composition of synthesis was: 27 $NH_2(CH_2)_6NH_2$:13.5 $K_2O:Al_2O_3$:90 SiO_2:3600 H_2O plus 0.71 wt% seed.

Recovery of the Solid Phase after Crystallization

After crystallization, the autoclave was removed from the oven and cooled to room temperature. The content was transferred to a beaker containing 100 ml of distilled water and the resulting solid was separated from the supernatant liquid by vacuum filtration, washed several times with distilled water (until the pH of the filtrate reaches the value ~7.0) and dried in an oven at 120°C for 12 h. In the case of the agitated synthesis (Parr reactor), the reactor vessel was cooled by immersion in an ice-water mixture to speed up the washing procedure and the cooling and recovering the solid phase was similar to that employed in the static synthesis.

Characterization

The samples were subjected to X-ray diffraction (XRD) Shimadzu, model XRD-6000. The XRD profile was used to identify the material (type of crystal structure), the calculation of the crystallinity of the material was performed using Equation (1), using the peaks of 2θ regions of 19.7 to 20.9 and from 23.6 to 25 degrees.

$$\text{Crist.}(\%) = \left(\frac{\sum \text{peak area of the sample}}{\sum \text{peak area of standard sample}} \right) \quad (1)$$

The verification of presence of contaminating phases was done by comparison with published data in the literature and the diameter of the crystals [9], was done based on the Scherrer Equation (2), where the peak was chosen in the range of 19.7 to 20.9 degrees 2θ.

$$D_{hkl} = \frac{K\lambda}{\beta \cos(\theta)} \quad (2)$$

The thermogravimetric analyses (ATG/DTG) were performed in the thermobalance Shimadzu model DTG-60H, with a heating rate 10°C/min in an atmosphere of synthetic air with a flow rate of 100 mL/min. From these analyses it was possible to quantify the organic template molecules and intracrystalline water present in the solid recovered from the crystallization process.

The absorption spectra in the infrared region (IR) were obtained with a spectrophotometer Fourier transforms infrared Shimadzu, Model IRPrestige-21. The spectra were obtained in the range 400 - 4000 cm^{-1} by using KBr as a dispersing agent.

RESULTS AND DISCUSSION

The initial phase of the study consisted of four syntheses varying the silicon source and keeping all other experimental parameters identical. The crystallization was carried out with stirring at 400 rpm, at 160°C for 24 hours. Figure 1 shows the X-ray diffraction (XRD) of the samples synthesized with colloidal silica and pyrolytic silica. From the comparison with the literature diffraction pattern [10] materials have been identified as zeolite ZSM-22 (TON structure).

The sample synthesized with colloidal silica had a higher crystallinity (~6%) when compared with the solid synthesized with pyrolytic silica.

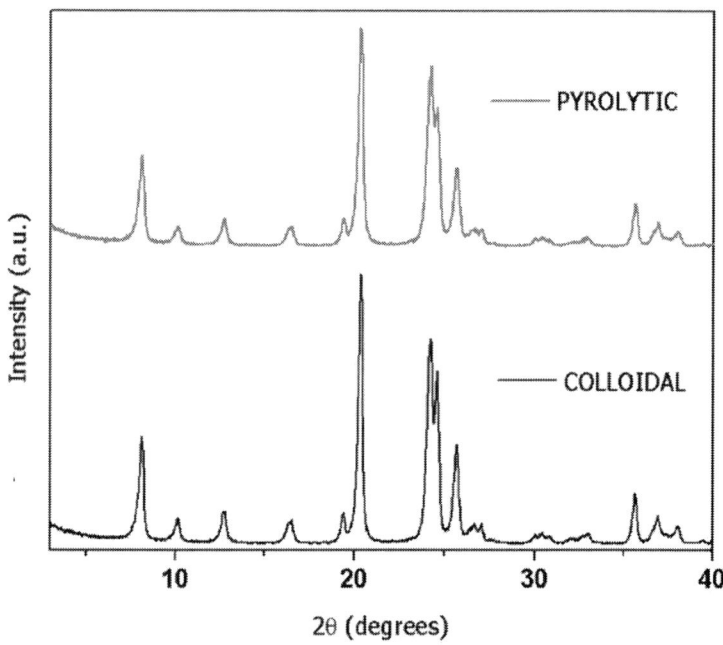

Figure 1: XRD of the samples crystallized under agitation employing colloidal silica and pyrolytic silica.

The samples with silica gel and cogel (compound solution resulting from the addition of aluminum sulfate + sulfuric acid diluted in a solution of commercial sodium silicate followed by filtration and washing) formed, respectively, the ZSM-5 zeolite and amorphous materials, as shown in Figure 2.

To verify the role of agitation on the crystallization of ZSM-22, a portion of the synthesis gel used in the experiments described above was transferred to a 70 mL autoclave and subjected to heating at 160°C for 24 hours under static conditions. The results of the analysis of X-ray diffraction indicated that for all sources of silica studied there was no formation of crystalline material (only an amorphous solid). A similar result was obtained by Derewinski & Machowska [11] for synthesis with colloidal silica utilizing the same organic template used in this study.

The sample of ZSM-22 synthesized with colloidal silica with the highest crystallinity was identified, and then grinded and treated by ultrasound for subsequent use as seed material.

The gel composition chosen to assess the role of the seeds was the same as the one used to generate the seed, i.e. a system with colloidal silica and the same amount of organic template. The crystallization conditions were similar to those described previously for static and agitated experiments. Figure 3 shows the XRD of the samples obtained with 24 h of synthesis.

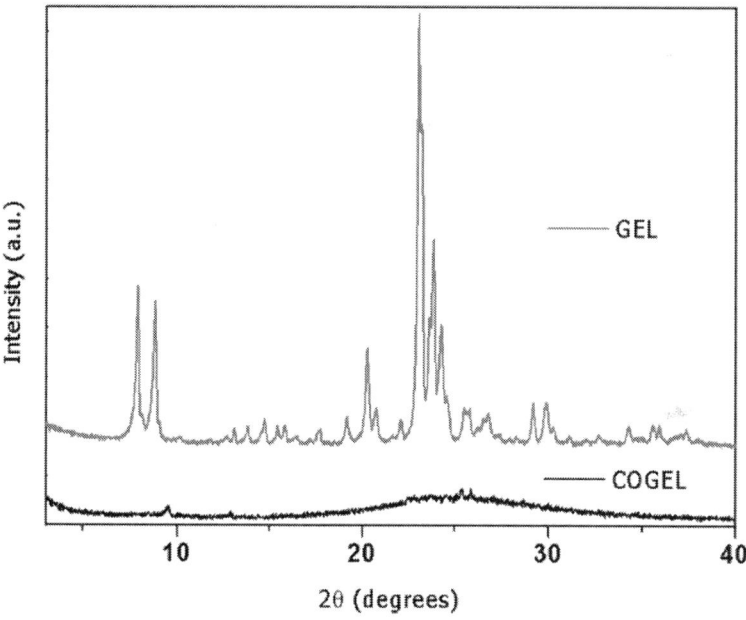

Figure 2: XRD of the samples crystallized with stirring using silica gel and the cogel obtained from sodium silicate.

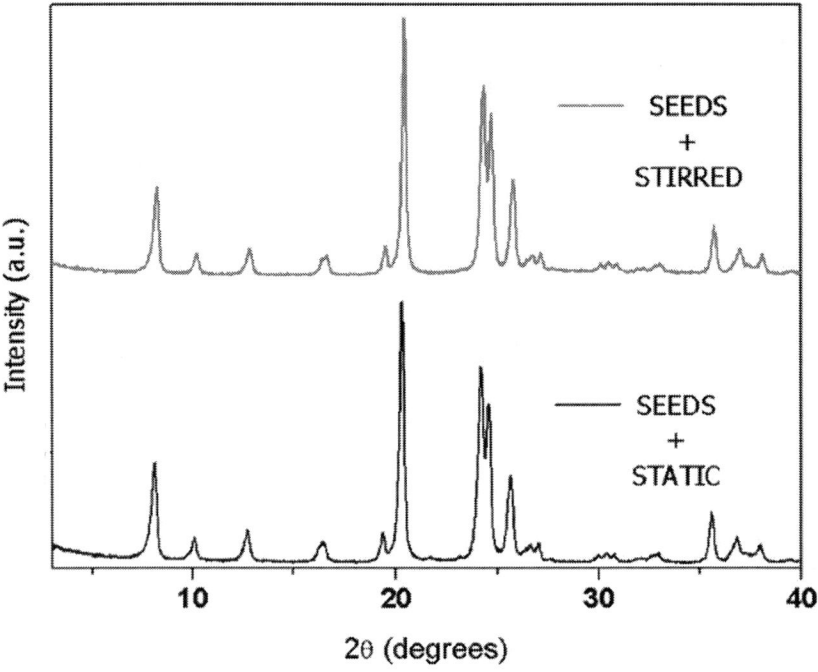

Figure 3: XRD of the ZSM-22 samples synthesized with seeds for static and agitated systems.

The experiment with seeds and agitation showed the highest degree of crystallinity (100% crystallinity of all the samples synthesized). Unlike the previous experiments without seeds, there was formation ZSM-22 in the system without agitation, but its percentage of crystallinity was slightly lower (~5%) compared to the agitated system.

In order to better understand the crystallization in static and agitated systems with addition of seeds, crystallization curves were constructed by periodic removal of samples, as shown in Figure 4.

In Figure 4, it is observed a rapid increase in the percentage of crystallinity of the samples from agitated systems—curves (a) and (c). It is noted that for the stirred seedless system the highest percentage of crystallinity was reached at 13 h. As for the stirred system with seeds, 100% of crystallinity was reached at 7 h. Therefore, the addition of seeds reduces by 4 hours the crystallization time to produce ZSM-22. In the case of static systems the addition of seed enables the formation of ZSM-22. However, the crystallization rate is slower than the one for stirred systems; the material is formed only after about 18h. This slower rate of crystallization is likely to be related to the difficulty in the diffusion of reactants within the reaction mixture to the surface of the growing crystal.

The average diameter of the ZSM-22 crystals was determined from the XRD patterns using the Scherrer equation, as indicated in Table 1. The data in the table indicates that crystals from seedless systems have smaller average diameters than the crystals obtained with seeds, as expected. This behavior may be due to two factors: 1) high speed stirring (400 rpm) prevents the formation of large crystals and 2) the use of seed introduces nuclei which speed the crystallization process resulting in larger crystals than for seedless systems. The calculated values for the average diameter of the crystals vary around 10%, because the calculation uses the peak areas from the XRD profiles where errors may surge from axial deviations, heterogeneity of the particles, preferential orientation, among others [12] [13].

The ATG/DTG curves of different samples of ZSM-22 synthesized in this work were very similar (Figure 5), having as the main feature four distinct events of mass loss.

These events were attributed to the following processes: 1) dehydration of the zeolite; 2) loss of organic template weakly bound to the surface of the material; 3) thermal decomposition of the organic template strongly bound to the surface; 4) loss of structural water by the condensation of silanol groups [4] [14] [15] and elimination of the fragments formed from the template decomposition.

The measurement of weight loss at each step was performed using a DTG curve and the data is summarized in Table 2. The mass loss events related to the removal of organic template (II and III) are responsible for about 2/3 of the total sample mass loss.

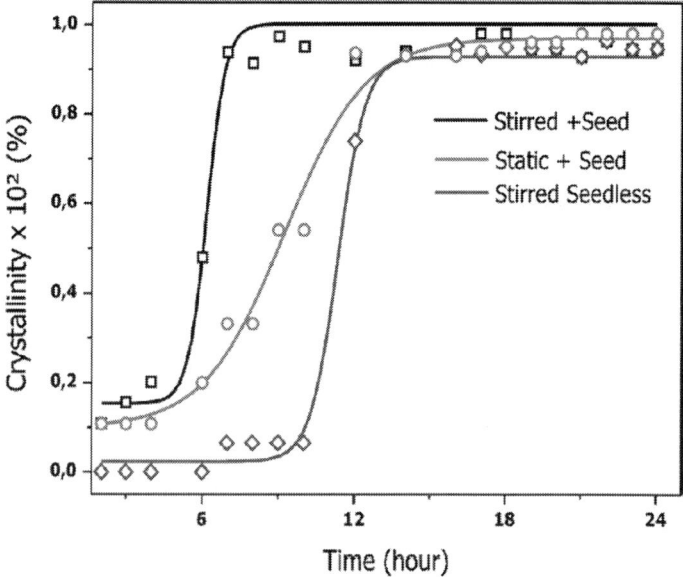

Figure 4: Crystallization curves. (a) Stirred + Seeds; (b) Static + Seeds; (c) Stirred Seedless.

Table 1: Average diameter of crystals obtained from the Scherrer equation for ZSM-22 samples with 24 hours of crystallization

Sample	Colloidal Silica	Pyrolytic Silica	Stirred + Seeds	Static + Seeds
Average Diameter of Crystals (nm)	32	28	40	39

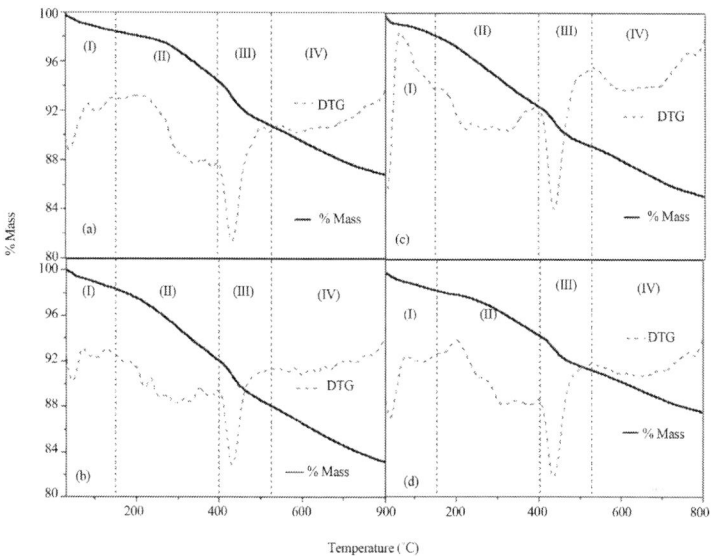

Figure 5: ATG/DTG curves (a) Colloidal silica; (b) Pyrolytic silica; (c) Seed + agitation; d) Seeds + static.

Table 2: Temperature ranges and mass losses obtained by ATG

Sample Events	Mass Loss (%)/Temperature (°C)				
	(I) 30 - 150	(II) 150 - 400	(III) 400 - 525	(IV) 525 - 800	Total Loss (%)
Colloidal sílica	1.36	4.01	3.57	3.96	13.98
Pyrolytic silica	1.66	6.32	3.94	4.94	16.87
Seeds + stirred	1.63	5.81	3.17	4.06	14.67
Seeds + static	1.59	4.02	3.04	3.71	12.35

Analysis of infrared spectroscopy for different ZSM-22 samples synthesized in this work is very similar. The spectrum for the most crystalline sample is shown in Figure 6. This curve shows five strong absorption bands in the region between 400 and 1500 cm^{-1}. The bands in at 1090 and 470 cm^{-1} correspond to the asymmetric stretching vibration type and flexion (T-O), for the internal tetrahedra. The band near 790 cm^{-1} is due to symmetrical stretching of the external tetrahedra [16]. In the range between 1500 and 4000 cm^{-1} (not shown) there are only two bands corresponding to large surface hydroxyl groups and intracrystalline water molecules. A striking feature of the spectrum is a shoulder band near 1220 cm^{-1} [17], which are typical of ordered silicates.

CONCLUSIONS

This work investigated the crystallization of ZSM-22 in agitated and static systems, and with or without the addition of seeds. It was found that it is possible to synthesize ZSM-22 zeolite free of impurities at 160°C in the presence of 1, 6-diaminohexane as the organic template using colloidal silica or pyrolytic silica as the silicon source only if the system is stirred during crystallization. The shortest crystallization time with the highest crystallinity for agitated systems without seed material is 13 h. The syntheses with silica gel or cogel (humid solid formed by amorphous aluminosilicate) even with agitation, do not result in ZSM-22. The use of silica gel formed ZSM-5, while the use of cogel resulted in an amorphous compound. The addition of seed material significantly accelerates the crystallization of ZSM-22, reducing the crystallization time to only 7 h, for agitated systems using colloidal silica. Furthermore, the use of seeds allows the formation of ZSM-22 with colloidal silica without the necessity of agitation during crystallization. The average diameter data indicates that agitation at 400 rpm during crystallization or using seeds (in the case of static systems) favor the rapid crystallization of ZSM-22 resulting in very small crystals in the range 28 - 40 nm. Thermogravimetric analyses of the ZSM-22 samples show four events of mass loss and a total loss in the range between 12% to 14%.

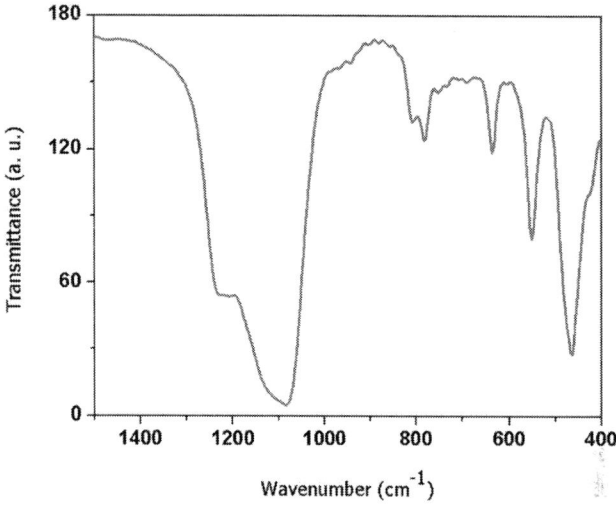

Figure 6: Infrared spectroscopy of ZSM-22 after 24 hours of crystallization without the addition of seeds with stirring.

ACKNOWLEDGEMENTS

The authors thank the Coordination of Improvement of Higher Education Personnel (CAPES), the Financier of Studies and Projects (FINEP), PETROBRAS (SINFER2 Project) and the National Council for Scientific and Technological Development (CNPq) and the Federal University of Alagoas (UFAL) for financial support and facilities to conduct this study.

REFERENCES

1. Ernst, S., Weitkamp, J., Martens, J.A. and Jacobs, P.A. (1989) Synthesis and Shape-Selective Properties of ZSM-22. Applied Catalysis, 48, 137.
2. Kumar, N., Lindfors, L.E. and Byggningsbacka, R. (1996) Synthesis and Characterization of H-ZSM-22, Zn-H-ZSM-22 and Ga-H-ZSM-22 Zeolite Catalysts and Their Catalytic Activity in the Aromatization of n-Butane. Applied Catalysis, 139, 189.

3. Kumar, R. and Ratnasamy, P. (1989) Somerization and Formation of Xylenes over ZSM-22 and ZSM-23 Zeolites. Journal of Catalysis, 116, 440.
4. Borade, R.B., Adnot, A. and Kaliaguine, S. (1991) Acid Sites in Al-ZSM-22 and Fe-ZSM-22. Zeolite, 11, 710.
5. Cundy, C.S. and Cox, P.A. (2005) The Hydrothermal Synthesis of Zeolites: Precursors, Intermediates and Reaction Mechanism. Microporous and Mesoporous Materials, 82, 78.
6. Thompson, R.W. and Robson, K.P.H. (2001) Nucleation, Growth, and Seeding in Zeolite Synthesis. In: Robson, H., Ed., Verifield Syntheses of Zeolitic Materials, Elsevier, Amsterdam, 21-23.
7. Verboekend, D., Chabaneix, A.M., Thomas, K., Gilson, J.P. and Ramírez, J.P. (2011) Mesoporous ZSM-22 Zeolite Obtained by Desilication: Peculiarities Associated with Crystal Morphology and Aluminium Distribution. Cryst. Eng. Comm., 13, 3408-3416.
8. Schmidt, W., Toktarev, A.V., Schuth, F., Lone, K.G. and Unger, K. (2001) Studies in Surface Science and Catalysis, 135-190.
9. Klug, H.P. and Alexander, L.E. (1954) X-Ray Diffraction Procedures. John Wiley & Sons, New York.
10. Treacy, M.M.J. and Higgins, J.B. (2007) Collection of Simulated XRD Powder Diffraction Patterns for Zeolites. 5th Edition, Elsevier, Amsterdam.
11. Derewinski, M. and Machowska, M. (2004) Effect of Stirring on the Selective Synthesis of MEL or TON Zeolites in the Presence of 1,8-Diaminooctane. In: van Steen, E., Callanan, L.H., & Claeys, M., Eds., Studies in Surface Science and Catalysis, Elsevier, Amsterdam, 349-354.
12. Connolly, J.R. (2013) Introduction to X-Ray Powder Diffraction. http://epswww.unm.edu/xrd/xrdclass/01-XRD-Intro.pdf
13. Connolly, J.R. (2013) Introduction Quantitative X-Ray Diffraction Methods.http://epswww.unm.edu/xrd/xrdclass/09-Quant-intro.pdf
14. Franklin, K.R. and Lowe, B.M. (1988) Hydrothermal Crystallization of (Hexane-1,6-Diamine)-Silicalite-1. Zeolite, 8, 495.
15. Singh, A.P. and Reddy, K.R. (1994) Synthesis, Characterization, and Catalytic Activity of Gallosilicate Analogs of Zeolite ZSM-22. Zeolite, 14, 290.

16. Silva, A.O.S. (2004) Sintese e Caracterização de Catalisadores de Ferro e Cobalto Suportados nas Zeólitas HZSM-12 e HZSM-5 para a Conversão de Gás de Síntese em Hidrocarbonetos. Universidade Federal do Rio Grande do Norte, Natal.
17. Simon, M.W., Suib, S.T. and Young, C.O. (1994) Synthesis an Characterization of ZSM-22 Zeolites and Their Catalytic Behavior in 1-Butene Isomerization Reactions. Journal of Catalysis, 147, 484.

Chapter 5

Nickel Alloy Catalysts for the Anode of a High Temperature PEM Direct Propane Fuel Cell

Shadi Vafaeyan[1,2] Alain St-Amant[3], and Marten Ternan[4]

[1]Chemical and Biological Engineering, University of Ottawa, Ottawa, ON, Canada K1N 6N5

[2]Centre for Catalysis Research and Innovation, University of Ottawa, Ottawa, ON, Canada K1N 6N5

[3]Department of Chemistry, University of Ottawa, Ottawa, ON, Canada K1N 6N5

[4]EnPross Inc., 147 Banning Road, Ottawa, ON, Canada K2L 1C5

ABSTRACT

High temperature polymer electrode membrane fuel cells that use hydrocarbon as the fuel have many theoretical advantages over those that use hydrogen. For example, nonprecious metal catalysts can replace platinum. In this work, two of the four propane fuel cell reactions, propane dehydrogenation and water dissociation, were examined using nickel alloy catalysts. The adsorption energies of both propane and water decreased as the Fe content of Ni/Fe alloys increased. In contrast, they both increased as the Cu content of Ni/Cu alloys increased. The activation energy for the dehydrogenation of propane (a nonpolar molecule) changed very little, even though the adsorption energy changed substantially as a function of alloy composition. In contrast, the activation energy for dissociation of water (a molecule that can be polarized) decreased markedly as the energy of adsorption decreased. The different relationship between activation energy and adsorption energy for propane dehydrogenation and water dissociation alloys was attributed to propane being a nonpolar molecule and water being a molecule that can be polarized.

INTRODUCTION

In principle, all types of fuel cells can convert the chemical energy of fuel into electrical energy more efficiently than competing technologies such as batteries or combustion processes. We have been investigating fuel cells that use hydrocarbons fuels (natural gas for urban areas and liquefied petroleum gas, LPG, for rural areas) that react directly at the anode of the fuel cell. Direct hydrocarbon fuel cells, DHFCs, have several advantages over hydrogen or methanol fuel cells. Theoretically, hydrocarbon fuels can be even more energy efficient than either hydrogen or methanol. Storage of liquid hydrocarbons is technologically less complex and less costly compared to that for hydrogen gas. The cost of delivering conventional electrical power in rural areas is about an order of magnitude greater than in urban areas, even though the price charged for electrical power is similar. Therefore, a greater capital cost for fuel cells can be justified for rural areas compared to urban areas. For that reason, we are investigating DHFCs that operate on propane fuel, because infrastructure for its delivery currently exists in many rural locations. Despite their many advantages, DHFCs have one substantial drawback. They have much smaller current densities (reaction rates) than fuel cells using hydrogen or methanol.

There are three reviews [1–3] that describe the extensive research done on DHFCs during the 1960s. Because that research did not produce the improvement in technology that was wanted, interest in the topic diminished, although research has continued to the present time [4–9]. Our group's strategy includes fuel cell reactor modeling using computational fluid dynamics, CFD [10–12], and fuel cell catalyst modeling using density functional theory (DFT) [13, 14]. More recently, we have also been performing experimental work.

Our research is directed toward the development of high temperature polymer electrolyte membrane, PEM, fuel cells operating near 120°C. In contrast, conventional PEM fuel cells operate near 80°C. One of the principal advantages of a 120°C operation is the elimination of liquid phase water. In the presence of water in the liquid phase, Pt, Ir, Au, and Pd are the only metals that are stable in such an acidic environment at +0.8 to +1.0 V [15]. The absence of liquid water corrosion from the accumulation of acidic species is diminished, so that the platinum group metals normally used as catalysts might be replaced by less expensive metals. Examples would be the catalysts used in alkaline fuel cells, nickel (at the anode), and silver (at the cathode). When the temperature of the Nafion electrolyte used in conventional PEM fuel cells is increased from 80 to 120°C, its proton conductivity decreases by an order of magnitude. Our experiments [16–18] suggest that a modified zirconium phosphate electrolyte supported in a porous polytetrafluoroethylene membrane may be appropriate for 120°C operation of direct hydrocarbon fuel cells.

The overall reaction in a direct propane fuel cell is shown in (1). The reaction at the anode is shown in (2), where EYL represents the electrolyte and me represents the solid transition metal electro catalyst that in this work is either a Ni-Fe alloy or a Ni-Cu alloy. The cathode reaction shown in (3) has been studied extensively by others [19–21] and is not part of this work. Consider the following:

$$C_3H_8 + 5O_2 = 3CO_2 + 4H_2O \qquad (1)$$

$$C_3H_8(g) + 6H_2O(g)$$
$$\longrightarrow 3CO_2(g) + 20H^+ (EYL) + 20e^- (Me) \quad (2)$$

$$5O_2 + 20H^+ (EYL) + 20e^- (Me) = 10H_2O \quad (3)$$

The anode reaction includes a single electrochemical reaction plus a multistage reaction network of chemical reactions (dehydrogenation of adsorbed C–H species, water dissociation, hydroxylation of adsorbed carbon species, and carbon-carbon bond cleavage) through which production of a number of intermediate by-products (such as propanol, propionaldehyde, propionic acid, ethanol, acetaldehyde, acetic acid, methanol, formaldehyde, formic acid, carbon monoxide, and carbon dioxide) is possible.

The initial steps of the chemical reaction network at the anode, (4) and (5), were the ones studied in this work. The single electrochemical reaction (6) that occurs is the ionization of hydrogen atoms where the hydrogen atoms can be formed via either (4) or (5). Of the 20 protons produced in the stoichiometric reaction at the anode (2), 8 come from the propane molecule and 12 are derived from water. Therefore, (5) is an important reaction.

$$C_3H_8 = C_3H_7 + H \quad (4)$$

$$H_2O = H + OH \quad (5)$$

$$H = H^+ + e^- \quad (6)$$

A number of DFT studies on various fuel cell reactions have been reported in the literature. Some examples are as follows: the oxygen electroreduction reaction [22], the hydrogen electrooxidation reaction in hydrogen fuel cells [23], the electroreduction reaction in the presence of liquid water [24], and the methanol electrooxidation reaction in methanol fuel cells [25]. The synergy between computational studies and experimental studies for fuel cells has been described recently [26].

The focus of this work has been on the use of metal alloys as anodic catalysts capable of the various reaction steps in the propane electrooxidation process. There are many studies investigating metal alloys as electrocatalysts in fuel cells, including the following topics: an anode material for low temperature fuel cells [27], carbon poisoning of cathode electrocatalyst alloys in SOFCs [28], the effect of the electrocatalyst on the oxygen reduction reaction [29, 30], and anodic methanol electrooxidation reaction on clusters of 2nd and 3rd row group VIII transition metals and Pt-Ru alloys [31].

The objective of this work was to investigate nickel metal and nickel alloys as possible anodic catalysts by performing DFT computations. Nickel, copper, and iron are 3D nonprecious metals. Various compositions of Ni/Fe and Ni/Cu alloys were examined for both the propane dehydrogenation reaction and the water dissociation reaction.

METHODS AND COMPUTATIONS

SIESTA software based on Kohn-Sham density functional theory was used to perform quantum chemical computations by Soler et al. [32]. The generalized gradient approximation (GGA) method was used as the exchange correlation functional type, with Becke-Lee-Yang-Parr (BLYP) parameterization. The default basis set, a double- polarization set composed of a compact and a diffused orbital basis, was used in these calculations. The tolerance of the density matrix was set to 10^{-3}. This value sets the maximum allowable difference between the output and the input on each element of the density matrix in a self-consistent field, SCF, cycle. A 4 × 4 × 4 (4*4*4/2 = 32k-points) Monkhorst-Pack k-point mesh was used. The convergence as a function of the number of -points was carefully monitored. Increasing the -point mesh from 32k-points to 48-kpoints changed the adsorption energies by an insignificant amount (~4 × 10^{-5} eV or ~8 × 10^{-3} kJ/mol). In some calculations, the coordinates of atoms were allowed to relax (change position) to determine the geometry having the minimum energy. When the difference in energy between successive calculations was less than 10^{-3} eV and when the maximum atomic force was less than 0.01 eV/Å, the convergence criteria were attained and the atomic geometry was optimized.

The pseudopotential for nickel was generated using the ATOM program of SIESTA. The Perdew-Burke-Ernzerhof (PBE) [33] exchange correlation was used for the generation of the electronic configuration. The improved Troullier Martins (tm2) method was used to generate the pseudopotential files for nickel, copper, and iron in their nonpolarized ground state electron configurations. The pseudopotential input file for a metal required that a particular core radius, r_c, be specified. It was obtained by trial and error. If the initial rc value was too small (≤1.5 for metals used in this study), the pseudopotential could not be generated. After a value of rc that was slightly greater than the minimum acceptable value had been obtained, the software would generate an exact value for r_c. That r_c value (3.34 Bohr for nickel, 3.18 Bohr for copper, and 3.64 Bohr for iron) was used to generate the pseudopotential. The pseudopotential output file was used for all subsequent computations that included that metal. To confirm the validity of our pseudopotential generation method and the resulting pseudopotentials, energy calculations were performed for a Ni unit cell using our custom generated Ni pseudopotential. The lattice constant (LC) that had the minimum total energy for the Ni unit cell was 0.370 nm. It differs by 4.9% from the experimental LC for Ni (0.352 nm) that was measured by X-ray diffraction, XRD; see Haynes [34]. This difference is smaller than the 6% error that was reported when using the PBE functional; see Perdew et al. [35].

DFT calculations were performed on a system of periodically repeated entities, defined in relation to nickel crystal unit cells. Each entity consisted of one propane molecule, a slab of nickel atoms, and a vacuum space. The slab consisted of two layers of atoms located on a x-y plane. The cells were arranged as follows: 3 cells × 3 cells with 4 atoms per cell (3 × 3 × 4 = 36 atoms per slab). The (100) surface was used for all nickel slabs. When calculations were performed with slabs containing Cu or Fe atoms, each Cu or Fe atom replaced a nickel atom in the original nickel slab. The vacuum space had a thickness of nine empty unit cells and was located above the nickel slab in the z direction. The vacuum space occupied the distance from the upper surface of a nickel slab within one periodic entity to the bottom surface of the next nickel slab above it in the next periodic entity. Ideally, the vacuum layer can prevent or minimize interactions between the periodic entities in the direction perpendicular to the surface. The choice of slabs with 36 nickel atoms per slab is typical of the number of metal atoms in many DFT slab calculations [36–39].

The slab surface should be large enough, in relation to the surface species to avoid both (a) interactions between the surface species on one slab and the nickel atoms in neighbouring slabs and (b) adsorbate-adsorbate interactions. Because there were 18 atoms in the surface of our slabs, the fractional surface coverage of propane was less than 0.2. It has been shown by Grabow et al. [40] that adsorbate-adsorbate interactions can influence the binding energy by as much as 1 eV when the fractional surface coverage increases from 0.2 to 1. Our low surface coverage minimized adsorbate-adsorbate contributions to the adsorption energy.

To confirm that the slabs were thick enough to predict propane adsorption on Ni (100), the thickness of the slabs was increased from 2 atoms to 4 atoms (increasing the number of atoms per slab from 36 to 72). The increased slab thickness only changed the value of the propane adsorption energy by 7.3% (~0.04 eV or ~4 kJ/mol). This relatively small variation indicates that calculations using slabs having two layers of atoms are sufficiently accurate to qualitatively predict trends in propane adsorption energy.

Another single adsorption energy calculation for propane was performed on a 2-layer Ni (100) slab in which the bottom layer was fixed at the positions of bulk nickel and the top layer was allowed to relax. When compared to the calculation in which the atoms in both layers were allowed to relax, the change in energy was also very small (4×10^{-4} eV ~0.04 kJ/mol).

The adsorption energies (E_{ads}) of the adsorbents (initially adsorbed reactants and subsequent adsorbed intermediate species) on the metal slabs were calculated according to the following equation:

$$E_{ads} = (E_{adsorbate+slab})_{MIN} - E_{species} - E_{slab}, \tag{7}$$

where $E_{adsorbate+slab}$ is the total energy for the adsorbate plus metal slab and $(E_{adsorbate+slab})$ MIN is the total energy for the configuration of the adsorbate plus metal slab that has the minimum energy, $E_{species}$ is the energy of an isolated reactant molecule in the gas phase, E_{slab} is the energy for a metal slab, and E_{ads} is the energy for adsorption of the species on the surface of the metal slab (or heat of adsorption).

The transition state energy, E_{TS}, for each reaction was obtained by calculating $E_{adsorbate+slab}$ as a function of the C–H bond length, the C–O bond length, or the O–H bond length. For each $E_{adsorbate+slab}$ calculation, the C–H bond length, the C–O bond length, or the O–H bond length was maintained constant, while all other bond lengths and all bond angles in the adsorbed species were relaxed to obtain the adsorbate configuration having the minimum energy for that particular C–H bond length, C–O bond length, or O–H bond length. Subsequently, the resulting energies that had been calculated at each C–H bond length, C–O bond length, or O–H bond length were compared as a function of their bond lengths. The configuration of the adsorbed species having the maximum energy, $(E_{adsorbate+slab})$ MAX, was the transition state (TS).

The transition state energies were obtained using the following equation:

$$\Delta E_{TS} = (E_{adsorbate+slab})_{MAX} - (E_{adsorbate+slab})_{MIN}. \tag{8}$$

Variations in the electrical potential of the slabs were not considered (zero electric field).

RESULTS AND DISCUSSION

Figure 1 shows the top view of the optimized structures of propane adsorbed on Ni/Fe alloy slabs. Propane is adsorbed in a somewhat similar manner on all of the Ni/Fe alloy slab surfaces. As the Ni/Fe atomic ratio decreases, the alignment of the three carbon atoms in the propane molecule becomes somewhat tilted with respect to the parallel rows of the metal atoms. The greatest tilt in alignment is on the surface of the pure Fe slab. Both Ni and Fe have fourfold symmetry but their crystal structures are different. Ni has a face-centered cubic, fcc, structure while Fe has a body-centered cubic, bcc, structure. In the first three images in Figure 1, the slabs have an fcc structure where the Fe atoms have replaced Ni atoms in that structure.

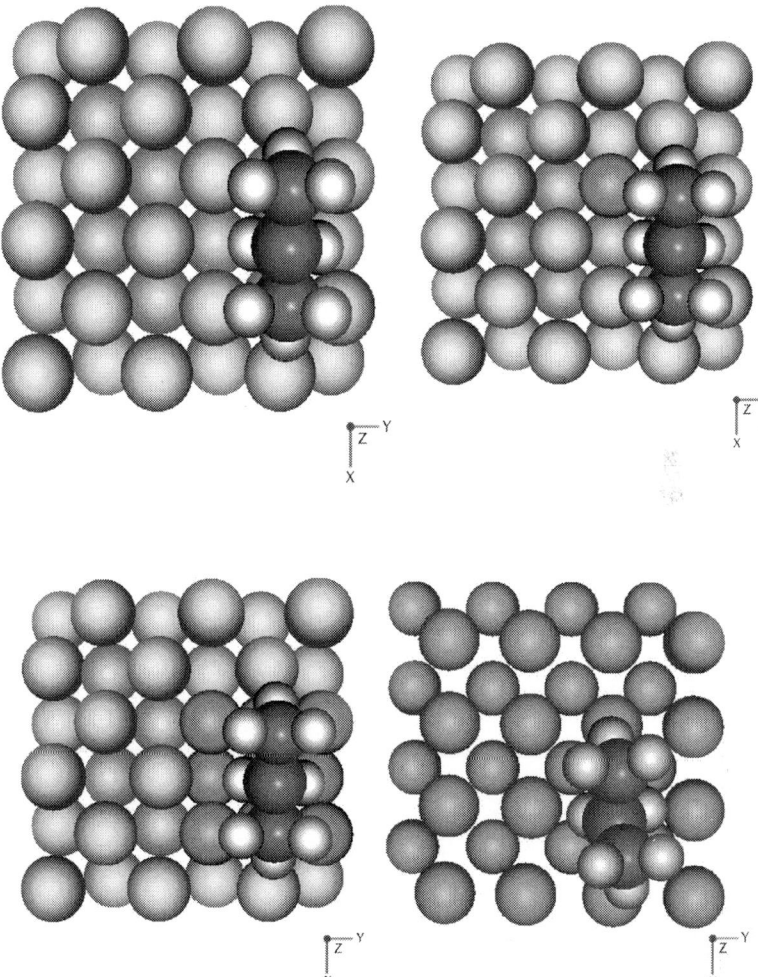

Figure 1: Propane adsorbed on the surface of Ni/Fe alloy anode slab (top view) with compositions of (a) 0% Fe (pure Ni slab), (b) 2.8% Fe, (c) 22.2% Fe, and (d) 100% Fe.

The propane adsorption energies are plotted in Figure 2 as a function of the Fe content in the slab. The energy of adsorption, E_{ads}, goes through a minimum as the Fe content of the slab increases. The diamond shaped data points represent alloys having ordered structures that have been defined. The diamond shaped datum point at 2.8% Fe is for a slab in which one of the Ni atoms has been replaced with a Fe atom that is adjacent to the Ni adsorption site, that is, the Ni metal atom on which the central carbon atom of the propane molecule is adsorbed. Even the presence of one Fe atom adjacent to the Ni adsorption site causes a small decrease in the adsorption energy. The diamond shaped datum point at 22.2 wt% Fe is for a slab in which the eight nearest neighbors of the Ni adsorption site have been replaced by Fe atoms. In this case, the decrease in adsorption energy is much more pronounced.

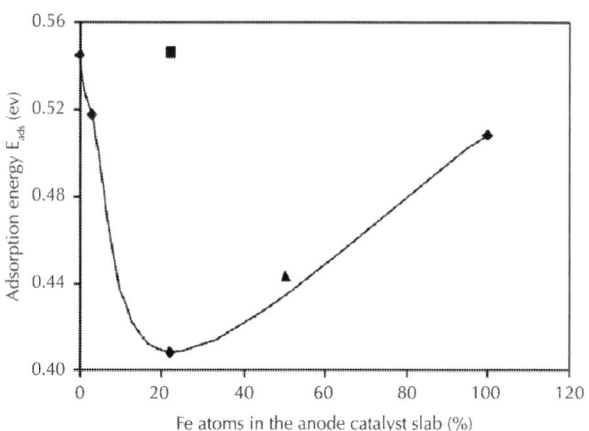

Figure 2: DFT calculated adsorption energies of propane on the surface of Ni/Fe alloy slabs with various compositions. The diamond shaped data points represent slabs having ordered structures. The triangular shaped datum point represents a slab containing 50% Ni and 50% Fe. The square shaped datum point represents a 22.2% Fe slab where the 8 Fe atoms have arbitrary locations and are further from the Ni adsorption site.

The square datum point represents a slab of 22.2 wt% Fe atom in which the 8 Fe atoms have arbitrary locations in the slab. In this case, most of the Fe atoms are not close to the Ni adsorption site. The adsorption energy is similar to that of the pure Ni slab. This suggests that an Fe atom needs to be adjacent to the Ni adsorption site if it is to influence the adsorption energy.

The triangular datum point represents a slab in which one-half of the Ni atoms in each layer were replaced by Fe atoms. However, the Fe locations were arbitrary. The ordered structures represented by the diamonds may be considered to be "designer" adsorption sites. In contrast, the result for a 50/50 Ni/Fe alloy with arbitrary Fe atoms locations suggests that a specific location may not be important as long as there are a sufficient number of Fe atoms in the slab. It also suggests that a specialized technique for preparing the alloy may not be necessary.

In Figure 3, replacing Ni atoms in the slab by Cu (with the same ratios as the Ni/Fe alloy slabs) causes a maximum in adsorption energy as opposed to the minimum observed in Figure 2. The opposite effect may be related to the number of d electrons in the metal. Copper has one more d electron than nickel. In contrast, iron has two fewer d electrons than nickel.

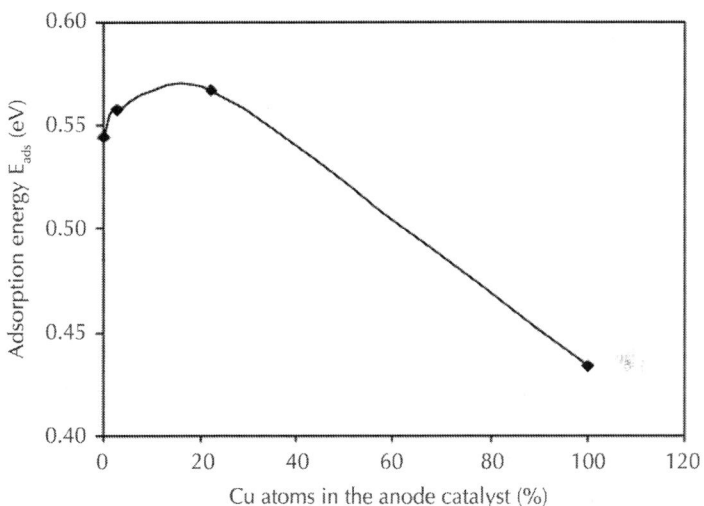

Figure 3: DFT calculated adsorption energies of propane on the surface of Ni/Cu alloy slab with various compositions (0% Cu, 2.8% Cu, 22.2% Cu, and 100% Cu).

The adsorption of water on four different Ni/Fe alloy slabs is shown in Figure 4. A slight change in orientation of the water molecule on the metal surface can be observed as the Fe content of the Ni/Fe alloy increases. The orientation of the water molecule on the Fe (100) surface is clearly different compared to that on the Ni(100) surface.

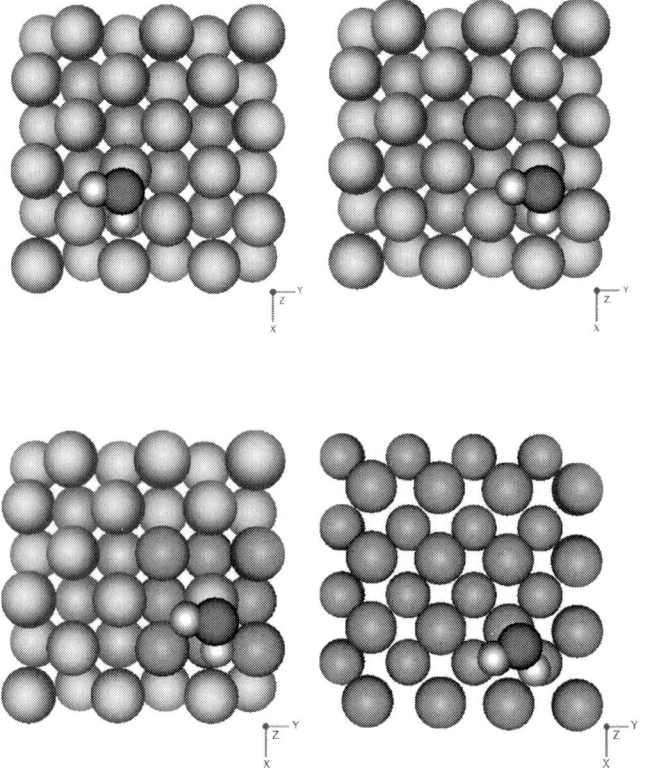

Figure 4: Water adsorbed on the surface of Ni/Fe alloy anode slab (top view) with compositions of (a) 0% Fe (pure Ni slab), (b) 2.8% Fe, (c) 22.2% Fe, and (d) 100% Fe.

The water adsorption energies are plotted in Figure 5 as a function of the Fe content in the Ni-Fe alloy slab. As the Fe content of the slab increases, the energy of adsorption, E_{ads} goes through a minimum. The shape of the plot is similar to the one for the adsorption of propane on Ni/Fe alloys.

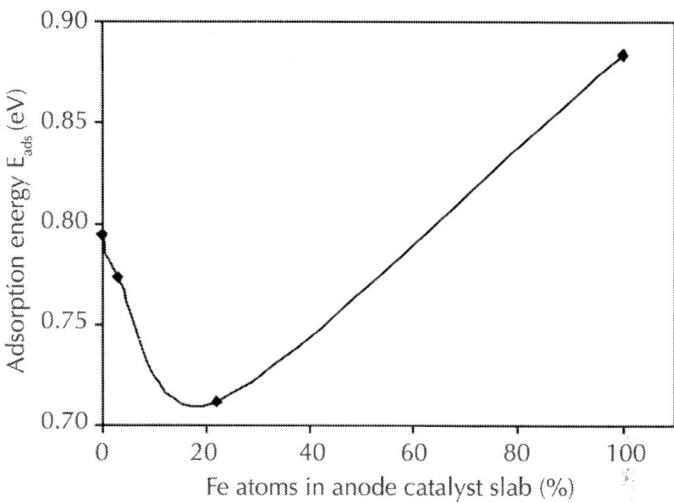

Figure 5: DFT calculated adsorption energies for water on the surface of NiFe alloy slab with various compositions (0% Fe, 2.8% Fe, 22.2% Fe, and 100% Fe).

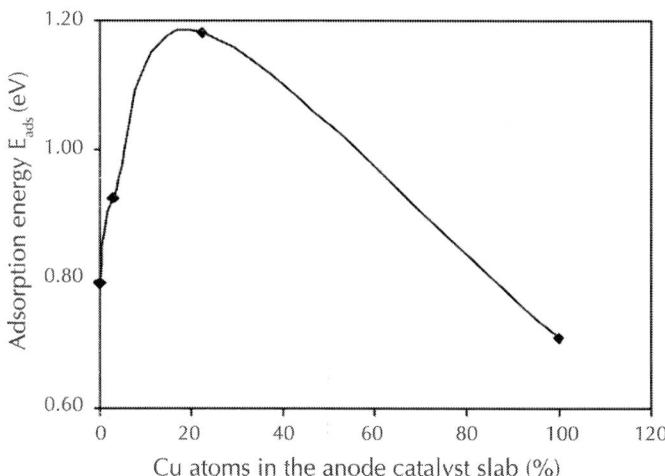

Figure 6: DFT calculated adsorption energies for water on the surface of Ni/Cu alloy slabs having various compositions (0% Cu, 2.8% Cu, 22.2% Cu, and 100% Cu).

The energy of adsorption appears to be a function of alloy composition. Both the propane and water adsorption plots have the same general shape for both Ni/Fe and Ni/Cu alloys. Changing the alloy composition from Ni to Ni/Fe or from Ni to Ni/Cu causes the same directional changes in energy of adsorption for both propane and water.

The activation energies for the dehydrogenation reaction of adsorbed propane to adsorbed propyl and atomic hydrogen species on Ni/Fe alloy slabs (with the same compositions used above for investigating the adsorption energies) were calculated using (5). The energy of a pure Ni slab plus its adsorbed species is shown in Figure 7 as a function of the C–H bond distance. It is the distance from the propane central carbon atom, adsorbed on the Ni surface to the hydrogen atom being removed during the dehydrogenation reaction. The activation energy (barrier height) is the difference between the energy of the adsorbed state (the minimum energy near C–H = 0.11 nm) and the transition state energy (the maximum energy near C–H = 0.17 nm). Similar calculations were performed to determine the activation energies for propane dehydrogenation and water dissociation for the various Ni/Fe and Ni/Cu alloys.

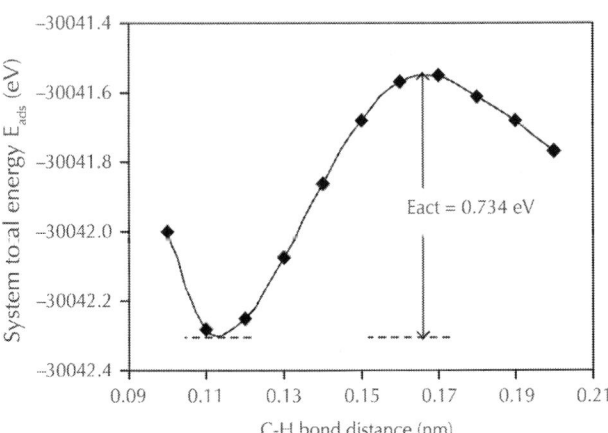

Figure 7: Energies of the adsorbate plus slab where the central carbon atom of the adsorbate had a defined C–H bond length. All the other bond lengths and all the bond angles in the adsorbate were permitted to relax to those values that produced the minimum energy.

The results for activation energy (energy barrier) calculations for propane dehydrogenation on slabs having different Ni/Fe compositions are shown in Figure 8. As the Fe content of the slab increases, the activation energy goes through a maximum. Since the smallest activation energies are obtained with either pure nickel or pure iron, it appears that Ni/Fe alloy catalysts are not helpful for propane dehydrogenation.

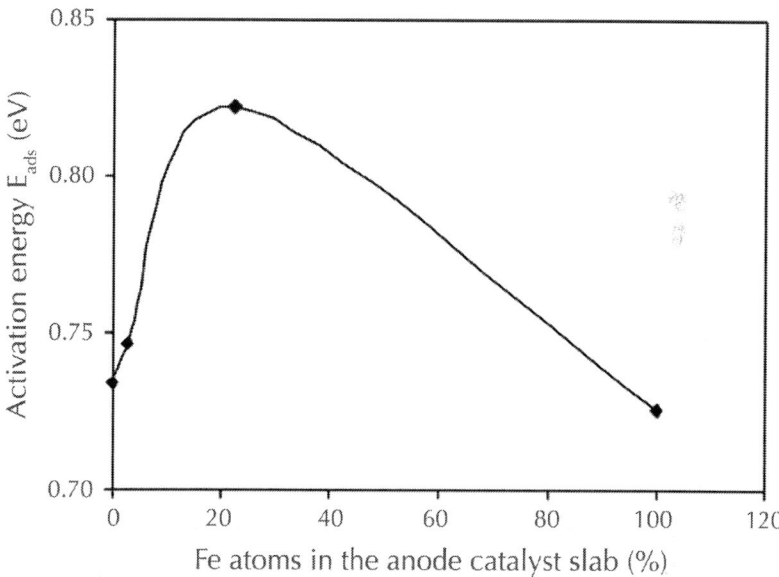

Figure 8: DFT calculated activation energy plot, E_{act} (eV), for the dehydrogenation of propane; $^*C_3H_8 \rightarrow {^*C_3H_7} + {^*H}$ (*indicating that the species is adsorbed on the surface of the catalyst) for various compositions of Ni/Fe anode catalyst slabs (0% Fe, 2.8% Fe, 22.2% Fe, and 100% Fe).

The activation energy for the water dissociation reaction is shown in Figure 9. There was a substantial decrease in activation energy as the Fe content in the Ni/Fe alloys increased. This indicates that Ni/Fe alloy catalysts will be superior to pure Ni or pure Fe catalysts for the water dissociation reaction. The combined data in Figures 8 and 9 suggest that an ideal catalyst might contain domains of pure metal for propane dehydrogenation adjacent to domains of Ni/Fe alloys for the water dissociation reaction.

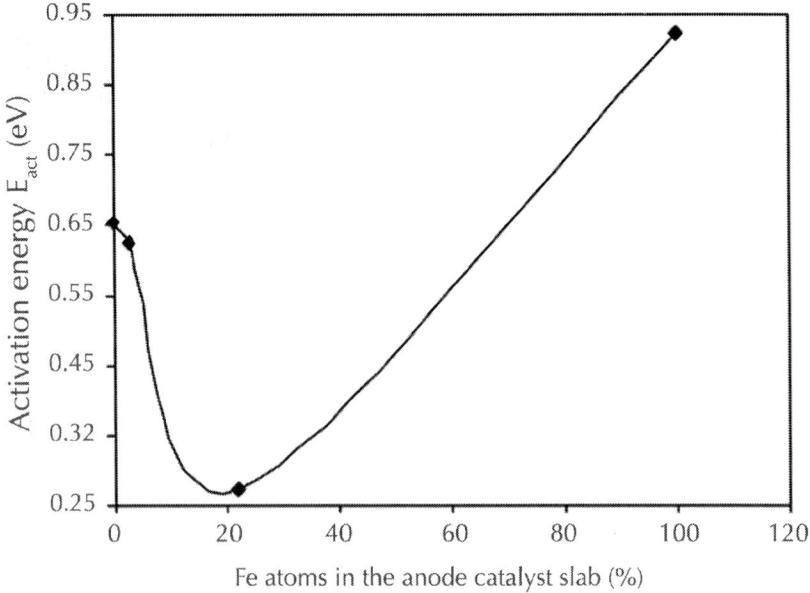

Figure 9: DFT calculated activation energy plot, E_{act} (eV), for water dissociation; *$H_2O \rightarrow$*OH + *H on Ni/Fe anode catalysts of various surface compositions (0% Fe, 2.8% Fe, 22.2% Fe, and 100% Fe).

Activation energies are plotted as a function of adsorption energies in Figure 10. For propane dehydrogenation, there is almost no change in activation energy. However, for water dissociation, the activation energy decreases substantially as the adsorption energy decreases. Propane is a nonpolar molecule. In contrast, water is a polar molecule. Perhaps the transition state formed from water may be more polarized. The difference in polarization of the molecules being adsorbed may be a part of the explanation for the different relationships between activation energy and adsorption energy.

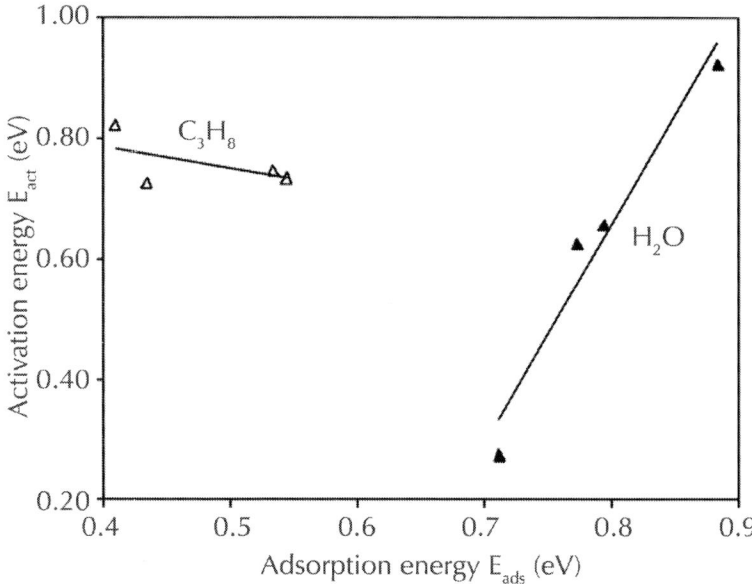

Figure 10: Activation energy, E_{act} (eV), versus adsorption energy, E_{ads} (eV), for propane dehydrogenation, "open triangles," and for water dissociation, "solid triangles."

The improvement obtained with Ni-Fe alloys reported here is consistent with two other reports in the literature. DFT computations [41] found that Ni-Fe alloys were superior to pure Ni for both CO and CO_2 hydrogenation reactions. Experiments [40] also showed that CO hydrogenation was more superior with Ni-Fe alloys than with pure Ni. It should be noted that both CO and CO_2 are polar molecules as is the H_2O molecule in this work. Secondly it is known that Pt-Ru alloys are superior to pure Pt [15] for the anode reaction in methanol fuel cells. Pt is in the same column of the periodic table as Ni and Ru is in the same column of the periodic table as Fe. Because methanol is a partially oxidized hydrocarbon molecule, the performance of direct methanol fuel cells might be expected to be similar in some ways to direct hydrocarbon fuel cells.

CONCLUSIONS

The activation energy or barrier height was determined for two reactions in direct hydrocarbon fuel cells. For the water dissociation reaction, a decrease in the activation energy was accompanied by a decrease in the energy of adsorption and by an increase in the Fe content of Ni/Fe alloys. In contrast, the activation energy increased when the Cu content of Ni/Cu alloys increased. For the propane dehydrogenation reaction, the activation energy was almost invariant with the energy of adsorption. The difference in the relationship between activation energy and adsorption energy was attributed to propane being a nonpolar molecule and water being a molecule that can be polarized.

ACKNOWLEDGMENTS

The authors are grateful for the financial support from the Canadian Government's Natural Sciences and Engineering Research Council and from the Ontario Provincial Government's Ministry of Research and Innovation (Ontario Fuel Cell Research and Innovation Network).

REFERENCES

1. H. A. Liebhafsky and E. J. Cairns, "The direct hydrocarbon fuel cell with aqueous electrolytes," in Fuel Cells and Fuel Batteries: A Guide to Their Research and Development, pp. 458–521, John Wiley & Sons, New York, NY, USA, 1968.
2. J. O. Bockris and S. Srinivasan, "The oxidation of unsaturated hydrocarbons on platinum," in Fuel Cells: Their Electrochemistry, pp. 391–402, McGraw-Hill, New York, NY, USA, 1969.
3. E. J. Cairns, "Anodic oxidation of hydrocarbons and the hydrocarbon fuel cell," Advances in Electrochemical Sciences and Engineering, vol. 8, pp. 337–392, 1971.
4. P. Heo, K. Ito, A. Tomita, and T. Hibino, "A proton-conducting fuel cell operating with hydrocarbon fuels," Angewandte Chemie—International Edition, vol. 47, no. 41, pp. 7841–7844, 2008.

5. E. P. Murray, T. Tsai, and S. A. Barnett, "A direct-methane fuel cell with a ceria-based anode," Nature, vol. 400, no. 6745, pp. 649–651, 1999.
6. O. Savadogo and F. J. Rodriguez Varela, "Low-temperature direct propane polymer electrolyte membranes fuel cell (DPFC)," Journal of New Materials for Electrochemical Systems, vol. 4, no. 2, pp. 93–97, 2001.
7. C. K. Cheng, J. L. Luo, K. T. Chuang, and A. R. Sanger, "Propane fuel cells using phosphoric-acid-doped polybenzimidazole membranes," Journal of Physical Chemistry B, vol. 109, no. 26, pp. 13036–13042, 2005.
8. F. J. R. Varela and O. Savadogo, "The effect of anode catalysts on the behavior of low temperature Direct Propane Polymer Electrolyte Fuel Cells (DPFC)," Journal of New Materials for Electrochemical Systems, vol. 9, no. 2, pp. 127–137, 2006.
9. T. Hibino, A. Hashimoto, T. Inoue, J. Tokuno, S. Yoshida, and M. Sano, "A low-operating-temperature solid oxide fuel cell in hydrocarbon-air mixtures," Science, vol. 288, no. 5473, pp. 2031–2033, 2000.
10. G. Psofogiannakis, Y. Bourgault, B. E. Conway, and M. Ternan, "Mathematical model for a direct propane phosphoric acid fuel cell," Journal of Applied Electrochemistry, vol. 36, no. 1, pp. 115–130, 2006.
11. H. Khakdaman, Y. Bourgault, and M. Ternan, "Direct propane fuel cell anode with interdigitated flow fields: two-dimensional model," Industrial and Engineering Chemistry Research, vol. 49, no. 3, pp. 1079–1085, 2010.
12. H. Khakdaman, Y. Bourgault, and M. Ternan, "Computational modeling of a direct propane fuel cell," Journal of Power Sources, vol. 196, no. 6, pp. 3186–3194, 2011.
13. G. Psofogiannakis, A. St-Amant, and M. Ternan, "Methane oxidation mechanism on Pt(111): a cluster model DFT study," Journal of Physical Chemistry B, vol. 110, no. 48, pp. 24593–24605, 2006.
14. S. Vafaeyan, A. St-Amant, and M. Ternan, "Propane fuel cells: selectivity for partial or complete oxidation," Journal of Fuels, vol. 2014, Article ID 485045, 9 pages, 2014.

15. J. N. Tiwari, R. N. Tiwari, G. Singh, and K. S. Kim, "Recent progress in the development of anode and cathode catalysts for direct methanol fuel cells," Nano Energy, vol. 2, no. 5, pp. 553–578, 2013.
16. A. Al-Othman, A. Y. Tremblay, W. Pell et al., "Zirconium phosphate as the proton conducting material in direct hydrocarbon polymer electrolyte membrane fuel cells operating above the boiling point of water," Journal of Power Sources, vol. 195, no. 9, pp. 2520–2525, 2010.
17. A. Al-Othman, A. Y. Tremblay, W. Pell, Y. Liu, B. A. Peppley, and M. Ternan, "The effect of glycerol on the conductivity of Nafion-free ZrP/PTFE composite membrane electrolytes for direct hydrocarbon fuel cells," Journal of Power Sources, vol. 199, pp. 14–21, 2012.
18. A. Al-Othman, A. Y. Tremblay, W. Pell et al., "A modified silicic acid (Si) and sulphuric acid (S)-ZrP/PTFE/glycerol composite membrane for high temperature direct hydrocarbon fuel cells," Journal of Power Sources, vol. 224, pp. 158–167, 2013.
19. Z. Liu, L. Ma, J. Zhang, K. Hongsirikarn, and J. G. Goodwin, "Pt alloy electrocatalysts for proton exchange membrane fuel cells: a review," Catalysis Reviews: Science and Engineering, vol. 55, no. 3, pp. 255–288, 2013.
20. D. C. Higgins and Z. Chen, "Recent progress in non-precious metal catalysts for PEM fuel cell applications," The Canadian Journal of Chemical Engineering, vol. 91, no. 12, pp. 1881–1895, 2013.
21. S. Stolbov and M. A. Ortigoza, "Rational design of competitive electrocatalysts for hydrogen fuel cells," Journal of Physical Chemistry Letters, vol. 3, no. 4, pp. 463–467, 2012.
22. F. Tian and A. B. Anderson, "Effective reversible potential, energy loss, and overpotential on platinum fuel cell cathodes," The Journal of Physical Chemistry C, vol. 115, no. 10, pp. 4076–4088, 2011.
23. E. Skulason, V. Tripkovic, M. E. Bjӧrketun et al., "Modeling the electrochemical hydrogen oxidation and evolution reactions on the basis of density functional theory calculations," The Journal of Physical Chemistry C, vol. 114, no. 42, pp. 18182–18197, 2010.

24. Y. Sha, T. H. Yu, B. V. Merinov, P. Shirvanian, and W. A. Goddard, "Oxygen hydration mechanism for the oxygen reduction 8 Journal of Chemistry reaction at Pt and Pd fuel cell catalysts," Journal of Physical Chemistry Letters, vol. 2, no. 6, pp. 572–576, 2011.
25. P. Ferrin and M. Mavrikakis, "Structure sensitivity of methanol electrooxidation on transition metals," Journal of the American Chemical Society, vol. 131, no. 40, pp. 14381–14389, 2009.
26. T. Zawodzinski Jr., A. Wieckowski, S. Mukerjee, and M. Neurock, "Integrated theoretical and experimental studies of fuel cell electrocatalysts," Electrochemical Society Interface, vol. 16, no. 2, pp. 37–41, 2007.
27. E. Christoffersen, P. Liu, A. Ruban, H. L. Skriver, and J. K. Nørskov, "Anode materials for low-temperature fuel cells: a density functional theory study," Journal of Catalysis, vol. 199, no. 1, pp. 123–131, 2001.
28. E. Nikolla, J. Schwank, and S. Linic, "Promotion of the longterm stability of reforming Ni catalysts by surface alloying," Journal of Catalysis, vol. 250, no. 1, pp. 85–93, 2007.
29. M. B. Vukmirovic, J. Zhang, K. Sasaki et al., "Platinum monolayer electrocatalysts for oxygen reduction," Electrochimica Acta, vol. 52, no. 6, pp. 2257–2263, 2007.
30. J. K. Nørskov, J. Rossmeisl, A. Logadottir et al., "Origin of the overpotential for oxygen reduction at a fuel-cell cathode," The Journal of Physical Chemistry B, vol. 108, no. 46, pp. 17886–17892, 2004.
31. J. Kua and W. A. Goddard III, "Oxidation of methanol on 2nd and 3rd row group VIII transition metals (Pt, Ir, Os, Pd, Rh, and Ru): application to direct methanol fuel cells," Journal of the American Chemical Society, vol. 121, no. 47, pp. 10928–10941, 1999.
32. J. M. Soler, E. Artacho, J. D. Gale et al., "The SIESTA method for ab initio order-N materials simulation," Journal of Physics Condensed Matter, vol. 14, no. 11, pp. 2745–2779, 2002.
33. J. P. Perdew, K. Burke, and M. Ernzerhof, "Generalized gradient approximation made simple," Physical Review Letters, vol. 77, no. 18, pp. 3865–3868, 1996.
34. W. M. Haynes, CRC Handbook of Chemistry and Physics, CRC Press, 91st edition, 2010.

35. J. P. Perdew, A. Ruzsinszky, G. I. Csonka et al., "Restoring the density gradient expansion for exchange in solids and surfaces," Physical Review Letters, vol. 100, Article ID 136406, 2008.
36. D. C. Ford, A. U. Nilekar, Y. Xu, and M. Mavrikakis, "Partial and complete reduction of O_2 by hydrogen on transition metal surfaces," Surface Science, vol. 604, no. 19-20, pp. 1565–1575, 2010.
37. T. Jiang, D. J. Mowbray, S. Dobrin et al., "Trends in CO oxidation rates for metal nanoparticles and close-packed, stepped, and kinked surfaces," The Journal of Physical Chemistry C, vol. 113, no. 24, pp. 10548–10553, 2009.
38. J. S. Hummelshoj, J. Blomqvist, S. Datta et al., "Communications: elementary oxygen electrode reactions in the aprotic Liair battery," The Journal of Chemical Physics, vol. 132, no. 7, Article ID 071101, 2010.
39. C. D. Taylor, M. Neurock, and J. R. Scully, "A first-principles model for hydrogen uptake promoted by sulfur on Ni(111)," Journal of the Electrochemical Society, vol. 158, no. 3, pp. F36–F44, 2011.
40. L. C. Grabow, B. Hvolbæk, and J. K. Nørskov, "Understanding trends in catalytic activity: the effect of adsorbate-adsorbate interactions for CO oxidation over transition metals," Topics in Catalysis, vol. 53, no. 5-6, pp. 298–310, 2010.
41. J. Sehested, K. E. Larsen, A. L. Kustov et al., "Discovery of technical methanation catalysts based on computational screening," Topics in Catalysis, vol. 45, no. 1–4, pp. 9–13, 2007.

Chapter 6

Role of Reaction and Factors of Carbon Nanotubes Growth in Chemical Vapour Decomposition Process Using Methane—A Highlight

Sivakumar VM[1], Abdul Rahman Mohamed[1], Ahmad Zuhairi Abdullah[1], and Siang-Piao Chai[2]

[1]School of Chemical Engineering, Engineering Campus, Universiti Sains Malaysia, Seri Ampangan, Nibong Tebal, Pulau Pinang 14300, Malaysia

[2]School of Engineering, Monash University, Jalan Lagoon Selatan, Bandar Sunway, Selangor 46150, Malaysia

ABSTRACT

One of the remarkable achievements in the field of nanotechnology is Carbon Nanotubes (CNT) synthesis. Since their discovery in 1991 by Iijima, CNTs have attracted much attention across the world. The CNTs are broadly classified into single-walled carbon nanotubes (SWNTs) and multiwalled carbon nanotubes (MWNTs). The most distinguished features of SWNTs and MWNTs are their electrical, mechanical, chemical, and electronic properties which in turn find their potential applications in almost all fields of science, engineering, and technology. Based on the previous research studies to till date, chemical vapour deposition (CVD) is considered to be the simplest method with high energy efficiency and precise control of reaction parameters compared to other different methods for synthesizing CNTs. Since production of CNTs is becoming the most important factor in the applications point of view, most industries today are opting for the CVD technique. This paper reviewed the synthesis of CNT by CVD especially focusing on methane CVD. Various parameters influencing the reaction and CNT growth were also discussed. A detailed review was made over the different types of CVD process, influence of metal, supports, metal-support interaction, effect of promoters, and reaction parameters role in CNTs growth.

INTRODUCTION

Carbon nanotubes (CNTs) are sheets of graphite rolled into tubes and possess excellent properties due to their symmetric structure [1]. They are broadly classified into single-walled carbon nanotubes (SWNTs) and multiwalled carbon nanotubes (MWNTs). Among them SWNTs are the key materials to the emerging field of nanotechnology [2]. CNTs have reached the forefront of many industrial research projects nowadays. Due to their high strength, stiffness, and electrical conductivity [3], CNTs are designated as one of the most attractive materials for reinforcing the material in composites [4, 5] and for nanoelectronics applications. Theoretical and experimental elastic modulus (1 TPa) and tensile strength of these materials are in the range of tens of GPa, respectively [6]. In general, CNTs can be produced by carbon arc discharge method (CA) [7], chemical vapour deposition method (CVD) [8], pulsed laser vaporization technique (PLV) [9], and high-pressure carbon monoxide conversion (HiPco) process [10]. In CA and PLV methods, although high quality materials can be produced, the high temperature employed for evaporating the carbon atoms from solid carbon sources (over 3000 K) make them difficult to scale up the process in a cost-effective way. Hence, chemical vapour decomposition (CVD) has gained importance owing to its easiest and economic way of production in a larger scale [11]. CVD method is believed as the most suitable synthesis method in terms of product quality and quantity [12]. A review by Baddour and Briens, 2005, concluded that catalytic technique such as CVD is simple, inexpensive, energy-efficient and can produce high purity CNTs in high yield (>75%) [13]. As the applications for CNTs range from nanoelectronics [14, 15], sensors [16, 17], and field emitters [18] to composites [19], reliable growth techniques capable of yielding high-purity material in desirable quantities are critical to realize CNT potential. This need is satisfied by CVD and the CNT growth factors. Since, there is a huge demand for CNTs production and application in the global market, it is necessary to maximize the yield and minimize the production cost. Hence, methane, which is found to be the most cheaply available resource from natural gas with high thermal stability and low Gibbs free energy, is suitable to be used. In addition, the current focus on SWNTs synthesis and its requirement of low carbon content source are fulfilled by methane. These properties and advantages paved way for the recent research on CVD focused towards using methane as hydrocarbon source.

CHEMICAL VAPOUR DEPOSITION

Catalytic chemical vapour decomposition (CCVD) was first introduced by José-Yacamán et al. in 1993 to produce CNTs [20]. A simple representation of the chemical vapour deposition process is shown in Figure1. CVD synthesis is achieved by putting carbon source in gas phase into a reactor and using an energy source, such as plasma or a resistively heated coil, to decompose the gaseous carbon molecule. In CCVD method, CNTs are produced by decomposition of carbon-containing molecules with the presence of catalytic materials. Commonly used carbon sources include methane, ethylene, hexane, ethanol, naphthalene, anthracene, carbon monoxide, carbon dioxide, acetylene, and benzene. Thermal or electrical energy source is used to crack the molecule into reactive atomic carbon. Then, the carbon diffuses into the supported metal, usually transition metal in Group VIII of the periodic table such as Ni, Fe, or Co. CNTs will be formed if the proper process conditions like reaction temperature, pressure, flow rate and concentration of hydrocarbon source, carrier gas, and so forth, are maintained. The appropriate metal catalyst among the transition group can grow SWNTs, DWNTs (double-walled carbon nanotubes), and MWNTs are shown in Figures 2(a), 2(b), and 2(c), respectively. Excellent alignment, size, diameter, growth rate as well as positional control on nanometer scale for the synthesized CNTs can be achieved by using CVD method.

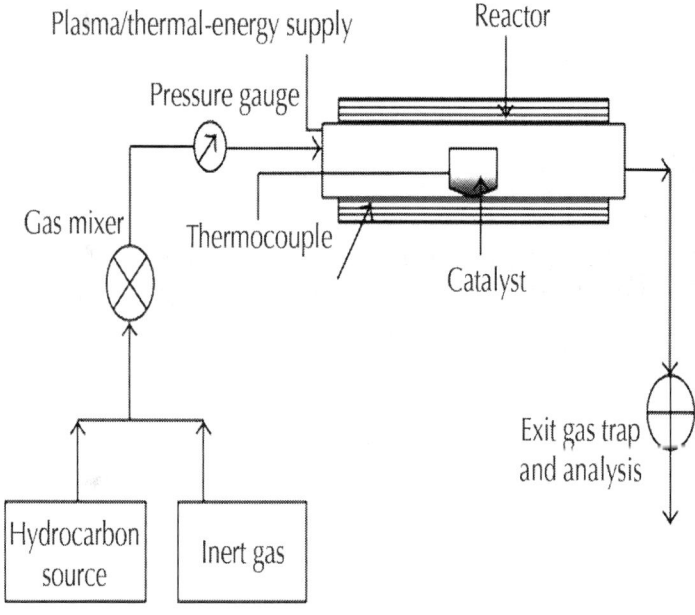

Figure 1: Schematic diagram of a CVD process for CNT synthesis.

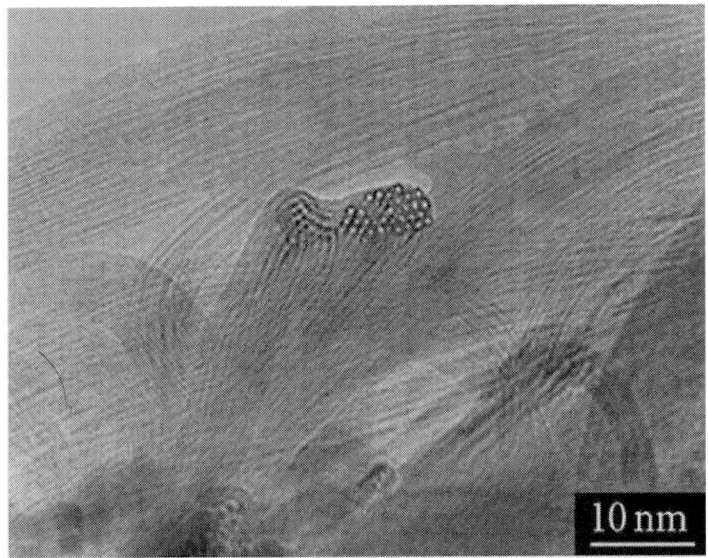

(a)

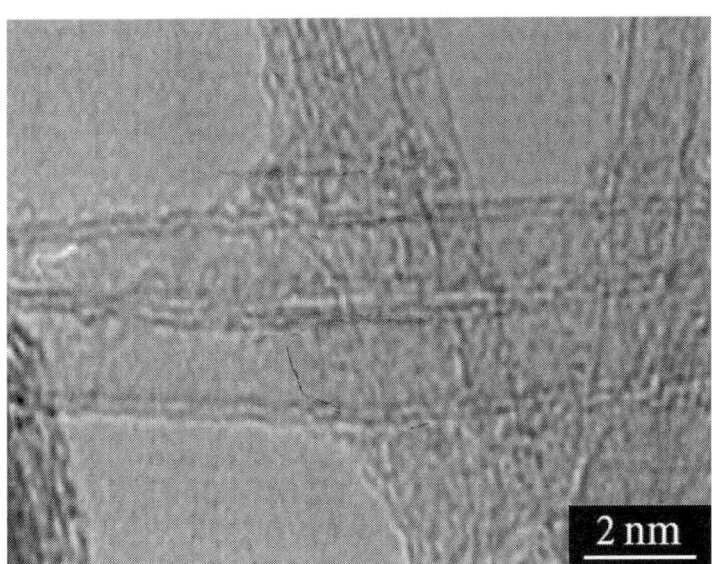

(b)

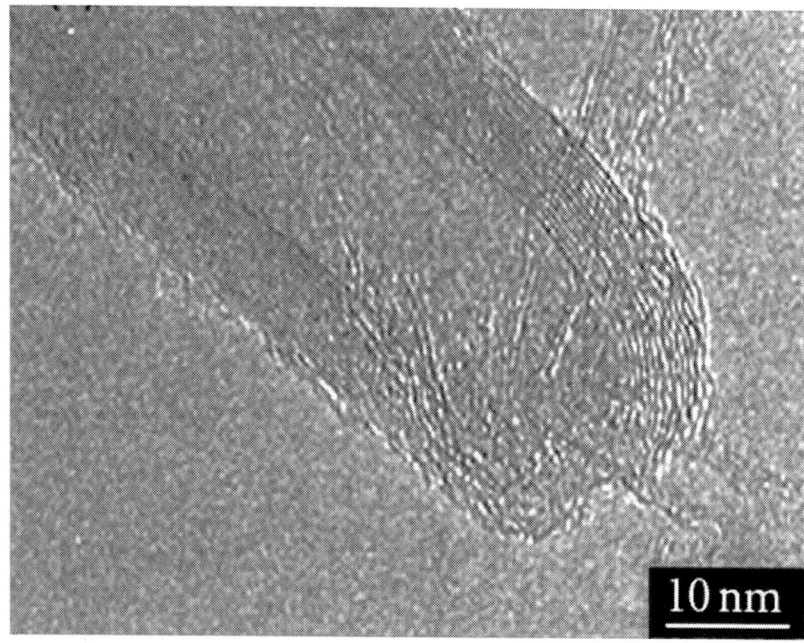

(c)

Figure 2: Transmission electron microscopy images (a) single-walled carbon nanotubes [21]; (b) double-walled carbon nanotubes [22]; (c) multiwalled carbon nanotubes [23].

Modified CVD Process

Nowadays the CVD process is under research with minor changes in their energy source to initiate the CNT growth. The process had been categorized according to their nature and source of energy (as shown in Figure 3). For examples, microwave [24, 25], inductively coupled plasma CVD [26], low pressure [27], hot filament (HF) [28, 29], alcohol catalytic [30], and so forth were reported.

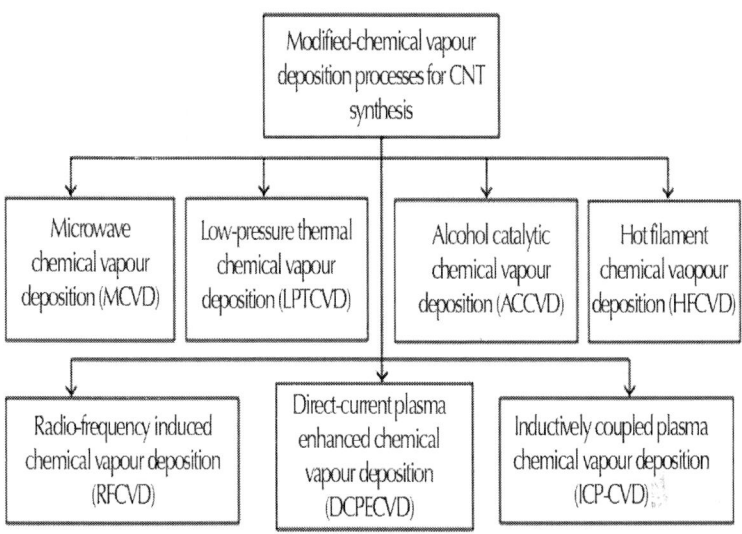

Figure 3: Classification of CVD processes according to their energy source.

Varadan and Xie [24], Fidalgo et al. [25] synthesized MWNTs by microwave CVD. The microwave system consists of a microwave magnetron with adjustable power supply ranging from 0 to 3000 W at a frequency of 2.45 GHz. No vacuum was maintained and reaction was operated at 1 atmosphere pressure. Straight and helical CNTs were obtained by decomposing acetylene in microwave energy field over the cobalt catalyst supported on Y-zeolite, alumina, and SiC at 700°C. High resolution transmission electron microscope (HRTEM) analysis results confirmed the presence of MWNTs. It was also reported that CNTs obtained using alumina support was found to have more impurities than zeolite. The main advantage of microwave CVD system is rapid heating and cooling process with a chance to produce helical CNTs with diameter range 80–100 nm.

Ikuno et al. [27] produced CNTs by low-pressure thermal chemical deposition (LPTCD) using pure ethylene. CNT bridges were grown on Ta electrodes at 800°C in vacuum. By nitriding the surface of the Ta electrode/SiO_2/Si substrates, Fe nanoparticles with a moderate size were effectively formed, resulting in bridging CNTs between the electrodes. It was found that under a low pressure of 100 Pa, straight CNTs are preferentially bridged between the Fe nanoparticles.

Yen et al. [26] synthesized well-aligned CNTs using a high inductively coupled plasma chemical vapor deposition (ICP-CVD) system. A gas mixture of CH_4–H_2 was used as the carbon source and Ni as the catalyst for the CNTs growth. The effect of process parameters, such as inductive RF power, DC bias voltage and ratio, on the growth characteristics of CNTs was investigated. It was found that generation and transport of ions to the substrate are the two underlying factors in determining the growth of CNTs.

Kadlečíková et al. [28] studied the effect of electric field upon the aligned growth of CNTs and had successfully produced bundles of CNTs by hot filament CVD method. The author stated that the key role in the formation of CNTs depends on the energy of ion bombardment and plasma discharge applied during the growth of CNTs after temperature and pressure were set in the deposition chamber.

Although various types of plasma enhanced CVD with specialized power supplies such as microwave, hot filament [28], radio-frequency, and direct-current (DC) exist and recognized as the most promising technique in growing CNTs at a relatively low temperature, this technique has a drawback of difficulty in scaling up the plasma technology to grow CNTs on a large scale and high voltages used lead to the sputtering of the electrode, causing both contamination of the plasma and damage to the CNT structure.

Unalan and Chhowalla [30] investigated SWNT growth parameters using alcohol catalytic chemical vapour deposition (ACCVD). Alcohol vapours like ethanol or methanol were used as the carbon sources and a mixture of Fe–Co acetate was used as catalytic material with MgO as its support. Dip-coating method was utilized for loading catalyst particles on the substrate. SWNTs with narrow diameter distribution without the presence of amorphous carbon were obtained at temperatures above 750°C, whereas defective MWNTs were observed at a lower temperature. It was reported that use of methanol as carbon source, with catalyst dissolved in deionised water (DI) rather than alcohol lead to growing more uniform SWNTs on the substrate surface.

In addition, a type of CCVD using fluidized bed reaction was also used for producing CNTs [30, 31]. Though the above techniques were tried with advancement in technologies, it still faces several drawbacks in scaling up the process. Moreover, the CCVD process is still found to be better for commercial synthesis of CNTs in proper controlled conditions [32, 33].

In the last decade, different techniques such as plasma enhanced CVD, thermal CVD, alcohol-catalytic (ACCVD) [34], vapour phase growth, and laser assisted CVD [35, 36] for the CNTs synthesis have been developed. Colomer et al. [37] confirmed that CCVD is easier, cheaper, and an adaptable approach for large-scale production CNTs at low temperature and ambient pressure via decomposing the hydrocarbon gas. Kong et al. [38] has already proved that CVD of methane can be a successful route for the growth of SWNTs.

Methane CVD

The following review has been made to highlight the current importance towards methane CVD in CNT synthesis using metal supported catalyst. High-temperature decomposition of hydrocarbons leads to the formation of deposited carbon. Morphology of the carbon deposits over the catalyst/support during CVD process was closely related to the thermodynamic properties and nature of the hydrocarbon source. Gaseous hydrocarbons such as acetylene [39–41], methane [42, 43], and ethylene [27] have been widely used for producing CNT deposits. Besides gaseous hydrocarbons, some liquid hydrocarbons such as hexane, benzene [44], toluene [45], xylene [46], and so forth, have also been effectively used for the growth of SWNTs and MWNTs in floating CVD methods.

Li et al. [47] studied the effect of various hydrocarbons like methane, hexane, cyclohexane, benzene, naphthalene, and anthracene over Fe catalyst impregnated on MgO support. The author reported that there was a strong dependence of nature of hydrocarbon precursors on the formation of different structure of CNTs. It was also reported that methane would be most preferable for the growth of high-purity SWNTs rather than any other gaseous hydrocarbons, as it was comparatively chemically stable at high temperature and has the simplest structure [48]. However, the achievement for the methane CVD growth of SWNTs would largely depend on the features of catalyst as well.

de Almeida et al. [49] studied the methane decomposition using porous-nickel alumina spheres as catalyst. It was observed that during catalytic decomposition reaction, methane is initially absorbed and decomposed on the metal surface of the catalyst particle, resulting in the formation of chemisorbed carbon species and the release of gaseous hydrogen. The carbon species was found to proceed further to dissolve in and diffuse through the bulk of the metal particle. Some of the aromatic hydrocarbons like benzene could be decomposed at a lower temperature when compared with methane [50]. But it was experimentally proven that benzene decomposition temperature should be kept at 800°C to favor the SWNTs formation, whereas in the case of methane, it could be achieved at a lower temperature of 650°C with lower Gibbs free energy ($\Delta G= -27.2$ KJ) [51]. SWNTs could not be obtained with hexane and cyclohexane though their reaction due to Gibbs free energies are much lower than that of benzene and methane. It was further revealed that the formation and morphology of carbon deposits were not simply determined by the pyrolytic behaviors of hydrocarbons.

Muradov [52] made a significant study of methane decomposition reaction (pressure: atmospheric, temperature: 850°C, methane flow rate: 5.0 ± 0.2 mL/min; sample amount: 0.3 ± 0.001 g) over 30 different samples of elemental carbon, including a variety of activated carbons (ACs), carbon blacks (CBs), nanostructured carbons (including CNTs and fullerenes), graphite, glassy carbon, and synthesized diamond powders. Catalytic activity of these carbons samples in methane decomposition was determined by their origin, structure, and surface area. ACs were found to provide the highest initial activity of methane decomposition with 90 vol% of H_2 formed for 10 s residence time with carbon sample bed. On the contrary, this highest methane decomposition rate with CO/CO_2-free hydrogen in a single step could not be obtained for experiments conducted with silica (surface area: 600 m^2/gm) and that of Ni- and Fe-based catalysts measured at identical conditions.

In 2007, Chuang et al. [53] developed sea urchin-like CNTs by catalytic decomposition of methane over catalyst-seeded mesoporous carbon and the catalysts were prepared by dip-coating method. The area density of CNTs/carbon nanowires (CNWs) was found to be higher at 900°C than that at 800°C. Most recently, Chuang et al. [53] and Fidalgo et al. [25] investigated the growth of nanofilaments on activated carbon and carbon fibre materials by microwave-assisted methane decomposition and they found that for pure methane flow at 800°C for 130 minutes, only amorphous carbon was formed over the activated carbon and carbon fibres.

It can be concluded that the nature of hydrocarbons plays a crucial role in CVD process for CNT formation, as well in economic aspects. It should be noted that the formation of SWNTs or MWNTs does not merely depend on carbon precursors, but also has strong correlations with other growth conditions. Since methane being a rich source from natural gas, several studies of CCVD using methane as the hydrocarbon source are under progress for investigating the other growth influencing conditions like nature of metal catalysts, support, metal-support interaction, their characteristics, temperature, pressure, gas flow rate, concentration, reaction time, and so forth.

CNT GROWTH PARAMETERS

Li et al. [23] studied that SWNTs or MWNTs formation does not alone depend on the nature of hydrocarbon source. There are also other important parameters playing more crucial role in the growth of different morphology of CNT. Based on distinguished CNT structure and chirality, their properties and end use applications are found to be significant. The main parameters for CNT growth are (i) properties, composition, and preparation method of metal catalyst, (ii) addition of promoter elements, (iii) metal-support interactions, (iv) reaction conditions like reaction temperature, pressure, inert/methane gas ratio, and gas flow rate.

Catalyst

Metals used to catalyze CNT formation are most often transition metals, in particular iron (Fe), cobalt (Co), and nickel (Ni), due to the solubility of carbon in these metals is finite [54]. It is noteworthy to find a large number of literatures reporting about CNT growth using different metals and their alloys [55]. The main reason of using transition metals is that they have nonfilled "d" shells and for that reason it is able to interact with hydrocarbons and show catalytic activity. In CVD process, transition metal particles act as a seed of nanotubes so that they strongly influence the structure and quality of the nanotubes. Though the transition metals can decompose hydrocarbons, they need a support for the growth of nanotubes. Materials like silica [56, 57], alumina [58, 59], zeolite [44], and recently MgO [60–62] were used as supports for active metals in developing different forms of carbon nanostructures like SWNT, DWNT, MWNT, and nanofibres. Different catalyst preparation techniques like impregnation [63–69], coprecipitation [70], sol-gel technique [71–73], thin film deposition using electron beam evaporation, and photolithography [74] for patterned catalyst on the support were studied. Among the above mentioned techniques, wet impregnation is the easiest way and mostly adopted by many of the researchers for preparing the catalyst.

Ichi-oka et al. [68] made a comparative study for the amount of carbon deposited via CVD of methane using nine different metals (Fe, Co, Ni, Ru, Rh, Pd, Os, Ir, and Pt) catalyst impregnated over MgO. It was reported that there was an increase in carbon content in the order of transition metal series: first < second < third row transition elements and the index of crystallinity (I_G/I_D) for CNTs in Raman bands decreased in the order: 8 > 9 >10 group elements in the periodic table.

Qian et al. [75] reported that higher conversion of methane into good morphology of CNTs was achieved over Fe and Mo catalyst. It was also reported that the metal particles with relatively small diameter distribution, high activity and thermal stability at high temperatures would meet the requirements of thermodynamics of methane conversion. Combined process of catalyst reduction and hydrocarbon decomposition were found to result with higher yield of CNT. Qian et al. [75] studied Co catalyst impregnated over Al_2O_3 support and reported 3-4 times of higher yield for combined process rather than separated methane decomposition in fluidized bed reactor.

He et al. [76] studied on the morphology of vertically aligned CNTs by using Fe on silicon substrates by CVD process. It was reported that decrease in the thickness of catalyst film reduced the diameter and increased the length of CNTs. He et al. [42] studied methane CVD over Ni/Al composite prepared by homogenous deposition and precipitation. The authors reported that MWNTs of 10–20 nm diameter range were formed from 20% wt Ni catalyst and carbon onions (5 to 50 nm) from 80% wt Ni catalyst. It was further found that the Ni/Al catalyst with 80 wt% Ni is less beneficial for CNT growth due to its increasing particle size which in turn leads to very low carbon solubility and resulted with carbon onions growth.

Zhu et al. [21] found that methane CVD over Co/MgO catalyst with the addition of Mo remarkably increased the yield of SWNTs by 10%

at least with suppressed amorphous carbon formation. It was reported that too many metallic Co particles with respect to Mo and support would form aggregated larger particles during preparation resulting with MWNTs.

Proper control of the catalyst particle size and metal concentration is critical for a high production rate of CNT. Yu et al. [77] studied the effect of Fe catalyst on silica particle size by CO disproportion method and revealed that particle sizes of 13–15 nm are optimum for maximum CNT growth rate. Therefore, small particle sizes do not always lead to a high growth rate. Li et al. [47] studied methane CVD for Co catalyst impregnated over MgO support. Co concentrations of 2.5–5 wt% were found to be efficient for growing DWNTs with diameters of 2–4 nm. It was reported that the growth was significantly influenced by catalyst concentration and type of supports. The important factors for a good catalyst are the control of the loading of the active metal and maintaining a good metal dispersion on the catalyst support [78].

Muradov [52] found that activated carbon (AC) catalyst played a significant role in methane conversion with initial H_2 concentration reaching up to 90 vol%. Also, it was stated that similar methane decomposition could not be achieved with silica gel (surface area: 600 m^2/gm) and that of Ni- & Fe-based catalysts measured at the identical conditions. The mechanism of methane decomposition over different forms of carbon as catalyst and their activity were yet to be explored in CCVD process for CNTs growth.

Metal-support Interaction

Catalyst supports are also reported to have great determination on the activity of a catalyst and the morphology of the produced CNTs. Metal-support interaction (MSI) is a very important factor in nanotube formation which is dependent on the nature of catalyst/support particle size, their surface area and the catalyst pretreatment process like calcination, reduction, and so forth. It has been found that CNT can be formed either with (i) tip-growth or (ii) base-growth model. Strong metal-support interaction of catalyst resulted CNTs to grow followed the base growth model, whereas tip-growth model is applied to CNTs grown by catalyst with weak MSI.

According to Vander Wal et al. [79] the formation of CNT structures is controlled by the effect of MSI. Though different sizes and shapes (helical, coiled, straight, and Y-shaped) of CNTs were produced so far, the growth mechanism of nanotubes is still a vague phenomenon for the researchers.

Several MSI studies for methane CVD have been carried out both in fluidized bed and fixed bed reactor system [80]. Catalytic performance of supported-NiO catalysts over methane decomposition at 550°C and 700°C by Chai et al. [81] shows that there is a decrease in activity of catalyst in the order of NiO/SiO_2>NiO/HZSM-5> NiO/CeO_2>NiO/Al_2O_3. From their XRD results, the dispersion of NiO particles on Al_2O_3 is found to be better compared to other support types. Also the formation of MWNTs at 550°C and SWNTs at 700°C is an indication of methane decomposition as well as the catalyst deactivation rate with respect to the increase in temperature. Similar studies on silica, zeolite, and alumina-supported Co catalysts for MWNT by acetylene decomposition show that percentage carbon deposition and diameter of CNT formed are based on the type and nature of support material [82]. The size of the pores on porous substrates determines the SWNTs or MWNTs formation. Substrate roughness studies by Ward et al. [11] also emphasize on the future research on substrate-catalyst interaction mechanism towards CNTs growth.

de Almeida et al. [49] quoted that methane decomposition at low temperatures (400–500°C) can be achieved using binary metal (Fe-M where M = Pd, Mo, or Ni) catalysts supported on alumina. Also, Muradov [52] studied methane decomposition over elemental carbon and reported that carbon materials are capable of producing H_2 rich gases at moderate temperatures. The author stated that catalytic activity of carbon was determined by their origin and surface properties. At higher temperature, such as at 830°C, $NiAl_2O_4$ phase is formed due to the strong metal-support interaction. Methane decomposition conversion values were found to increase with increasing reaction temperature. Hence, it can be concluded that the lower calcination temperature gave the catalyst with weak MSI but increased the methane conversion. In the case of higher calcination temperatures, Ni was incorporated strongly with alumina and thus methane conversion was low. Ward et al. [11] studied CNT synthesis on various substrates (alumina, silica carbide, silicon, quartz, sapphire, MgO, porous silicon, aerogel, fused silica, etc.) and its effect with Al-Fe-Mo by multilayer thin films deposition (electron beam evaporation). They found that Fe thin films spun on alumina to be the best for growing SWNTs and all other supports would give the formation of MWNTs primarily. Authors reported that the property of the substrates/supports like amount of crystallinity, pore size, and surface roughness had a major effect in the growth of either SWNTs or MWNTs

Chai et al. [83] studied the effect of CuO, FeO, and NiO added onto CoO/Al_2O_3 for methane decomposition. It was found that addition of these metal oxides shortened the catalyst lifetime and decreased the methane conversion. Catalyst lifetime and carbon capacity were found to increase with MoO-CoO/Al_2O_3. The carbon capacity over the supported CoO catalyst was found to decrease in the order of silica > zeolite > alumina > ceria > titania > magnesia > calcium oxide and the best catalytic carbon capacity was observed over CoO/SiO_2.

According to various researches, it was found that catalyst with weak MSI is efficient in methane decomposition whereas, for a catalyst with strong MSI, the metal particles are not easily detached from the support. Moreover, strong interaction increased the surface carbon accumulation over catalyst and enhanced the catalyst deactivation in methane decomposition.

Promoters

Some of the metals like molybdenum, boron [84], and sulphur [85] have been used as promoters for catalyst towards the decomposition of hydrocarbon gas into Y-shaped and helical CNTs. Molybdenum was found to be an effective dispersive agent for the catalyst when it is present at the optimum composition in the catalysts as reported by Yu et al. [62]. Ago et al. [86] found that the catalytic activity increases in the following order Fe-Mo > Fe > Co > Ni. The addition of molybdenum to iron catalyst increases the initial methane conversion and prevents the rapid deactivation of the catalyst.

Li et al. [47] studied the effect of Mo on MgO support. The formation of magnesium molybdate ($MgMoO_4$) phase was found to play a main role in MWNT synthesis. They suggested that ratio of Mo/Mg should be <1 for the growth of MWNT bundles. It was found that an appropriate amount of water vapour on the catalytic nanotube growth increases the CNT yield by 35%. Zhu et al. [21] showed that addition of molybdenum

(Mo/Co = 1/5) in an optimum level to Co/MgO catalysts (prepared by mechanical mixing and combustion synthesis with citric acid) yielded 10 time's higher generation of SWNTs and less amorphous carbon. Liu et al. [48] studied the CCVD of methane over Fe-Mo/Al$_2$O$_3$ catalyst. They reported that the yield of 70% SWNTs and 30% DWNTs are related to the weight of Fe-Mo metal in Fe-Mo/Al$_2$O$_3$. Similarly, Kang et al. [87] synthesized SWNTs and DWNTs over Fe-Mo/MgO catalyst and reported that proportion of DWNTs increased with an increase in reaction temperature. Moreover, Mo was known to be a catalytic centre for promoting the aromatization of methane. Zhang et al. [78] synthesized DWNTs by methane CVD on Fe/Al/MgO catalyst. The author found that introduction of Al species into Fe/MgO would reduce the size of MgO crystallites, providing a better dispersion of the metal particles on the support.

Sulfur was introduced as an additive in methane CVD process for the production of CNTs over Co-Mo/MgO catalyst prepared with sol-gel method [85]. It was found that the use of sulfur compounds during the sol-gel catalyst preparation process led to a significant change in the matrix composition and matrix-catalyst interaction. In this case, enhanced growth of Helical (HCNTs) was reported. Similarly, Y-shaped CNTs (YCNTs) growth was favored when sulfur in the form of thiophene vapours was used over the sol-gel prepared Co-Mo/MgO catalyst.

Phosphorous was reported in the literature by Ci et al. [88] for promoting effect in the formation of carbon filaments. Addition of phosphorous from a solution of H$_3$PO$_4$ in ethanol to the substrate, and followed by impregnation with Fe$_3$(CO)$_{12}$ was found to be effective in promoting the growth of vapour grown carbon fibres (VGCFs) [89]. Also, it was found that an increase in the amount of Phosphorous/Fe ratio >0.25 had an inhibiting effect on the VGCFs. He et al. [59] synthesized binary and triple CNTs over Ni/Cu/Al$_2$O$_3$ in the ratio 2:1:1 (prepared by sol-gel method) using −20 mesh to +40 mesh particles size for methane CVD at 550°C and the findings show that copper element provoked the formation of CNT.

REACTION PARAMETERS

The impact of reaction parameters like reaction temperature, reaction time, concentration of H_2, and flow rate ratio ($CH_4:H_2$) plays a major role in deciding the types of CNTs formation, and its yield. Recently, studies on effect of aforementioned reaction parameters in CNT formation using methane CVD in presence of Ni-Mo/MgO were reported by Zhan et al. [90]. Morphological structure of CNTs was also found to depend much on the reaction temperatures [91], reaction time [92], and CH_4 to H_2 gas ratio [93] in methane CVD.

Yu et al. [62] studied the influence of reaction atmosphere on Fe/MgO and Fe-Mo/MgO catalyst in argon and nitrogen by methane CVD. It was reported that there was an increase in diameter of SWNTs, MWNTs, and CNFs formed as the reaction was enriched in nitrogen atmosphere. On the other hand, pure SWNTs were obtained with Ar atmosphere. Ni et al. [61] studied the kinetics of CNT synthesis by methane CVD over Mo/Co/MgO and Co/MgO catalyst. It was found, the rate of CNT synthesis is proportional to the methane pressure, indicating that the dissociation of methane is the rate-determining step for a catalyst.

Zhao et al. [94, 95] claimed that reaction time would influence the morphology and diameter of CNTs. Carbon onions and two kinds of herringbone CNFs were obtained with 5 wt% Ni coprecipitated aluminium matrix. The catalysts used were calcined at 240°C and 400°C for 2 hours. Decomposition reaction was conducted in horizontal quartz boat at 550°C/600°C for 1 hour and 2 hours by varying the ratio of $CH_4:N_2$ from 1:4 to 1:5.5. The findings confirmed that carrier gas (N_2) plays an important role in CNT synthesis. The authors reported that cone-shaped catalyst was formed at 550°C and cylinder shape catalyst at 600°C. Influence of temperature over the shape and size of the catalyst was reported and thus the morphology of herringbone CNFs. Hence, by extending the growth time from 1 to 2 hours, carbon onions were formed.

Noda et al. [96] quoted various studies on methane decomposition over different catalyst and supports for the formation of SWNTs and MWNTs at different reaction temperatures. Effect of 1% and 5% of Ni loaded on SiO_2 in the temperature range of 625–800°C were studied in a fixed bed quartz tube reactor. Catalyst with 1% Ni favored

SWNT formation at all reaction temperatures whereas 5% Ni formed MWNTs at low temperature and SWNT at high temperature. Also, it was found that the amount of CNT decreased at high temperature for both concentrations due to sintering process.

de Almeida et al. [49] studied CH_4 deposition by preheating Ni/Al_2O_3 samples at 350,550 and 700°C under air flow for 1 hour and reaction at atmospheric pressure with molar ratio of $N_2:CH_4 = 6:1$ at 600°C, 700°C, and 800°C. He reported that higher calcination temperature lowers the residual amorphous carbon present within the pores of crystalline catalyst sample, resulting in mesopore formation and increasing surface area and pore volume.

Qian et al. [31] investigated the combination of catalyst reduction and methane decomposition over Ni and Co in fluidized bed reactor. Co/Mo/Al_2O_3 of ratio 1/4/50 wt ratio and Ni/Cu/Al_2O_3 of 15/3/2 wt were prepared by coprecipitation and decomposed in the range of 550°C–850°C with methane to argon ratio of 1:1. Higher methane conversion and 3-4 times better yield of CNTs were reported for this combination. He states that when the metal catalysts are reduced, they provide energy for the endothermic methane decomposition and make methane decomposition equilibrium shift to the direction of hydrogen & carbon and CNTs production.

Bustero et al. [97] studied the effect of temperature, mass of catalyst, initial feed gas composition, and reaction time on the yield of CNT. Optimal operating conditions for the synthesis of CNTs by CVD as reported by the authors are the reaction time: 10 min; temperature: 1000°C; catalyst Mass: 0.5 g; ratio of H_2/CH_4 is 1:1. A mathematical expression was established between the processing condition and the yield of carbon deposits. The derived expression also confirms that It is difficult to determine the optimal reaction conditions for the CNT synthesis by CVD.

Ni et al. [61] studied the effect of pressure and temperature on CNT synthesis by methane CVD over Mo/Co/MgO and Co/MgO catalyst. An increase in carbon yield was observed for an increase in methane pressure from 7.5 Torr to 78.0 Torr. Even at high methane pressures, no significant deactivation of the Mo/Co/MgO catalysts was noted. It was found that the rate of methane disassociation was reduced due to the addition of Mo into Co/MgO catalysts.

In 2007, Chuang et al. [53] conducted catalytic methane CVD over metal catalyst supported over mesoporous carbon (prepared by dip-coating method) at different temperatures (800°C and 900°C). It was concluded that at 900°C, CNTs and carbon nanowires (CNWs) could be obtained in high density rather than at 800°C. Recently, Fidalgo et al. [25] studied the influence of different CH_4/N_2 ratio with respect to nanofilaments formation. Methane CVD was conducted over Fe and Ni catalyst impregnated over activated carbon and carbon fibres at 800°C for 130 minutes. The growth of nanofilaments was found to be more abundant using $CH_4:N_2$ ratio of 1:3 rather than 1:1. It is concluded that the presence of N_2 in methane CVD could influence the carbon yield. The factors of CNT growth have been summarized in the Figure 4.

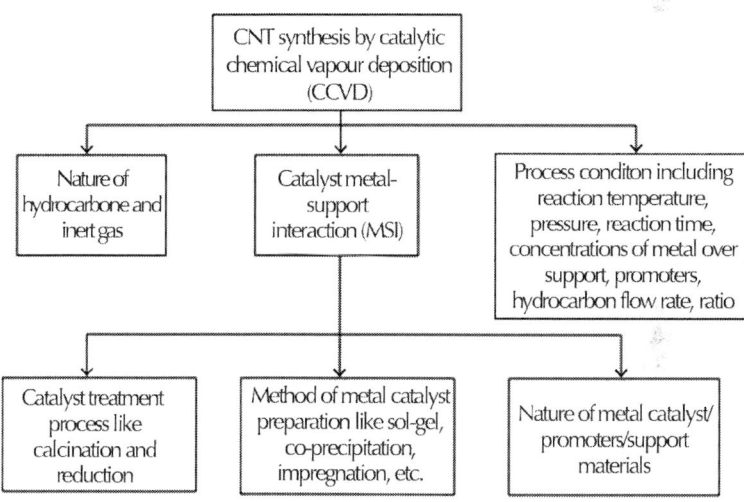

Figure 4: The network of the important factors for CNT synthesis.

It was found that methane CVD processes had been influenced by the process and CNT growth parameters like methane/carrier gas ratio, reduction atmosphere, methane/inert gas volumetric flow rate, reaction temperature, time of reaction and operating pressure, and so forth, Hence all the above-mentioned parameters would be crucial in deciding the nature, properties, yield, and quality of CNTs.

CONCLUSIONS

The current focus on the CCVD of methane for CNTs synthesis was reviewed. Methane being a highly available hydrocarbon source with high thermal stability and thermodynamic properties as discussed, still needs to be studied with the other influencing parameters on CNT growth. The carbon-based supports like AC is found to be a good material in decomposing methane at low temperatures. As mentioned by various researchers, each step starting from the metal catalyst preparation to removal of impurities from the synthesized CNTs would reflect in its product type, quality, yield, and ultimately its market demand for their potential applications. In the economic aspects, the value of CNT material, which at present greater than the value of gold, can be made available in bulk by proper control and optimization of the reaction parameters. Growth mechanism, which is still a vague phenomenon for many researchers, needs further investigations. Method of purification, a deciding factor for the purity of CNTs, needs to be identified to selectively separate the material based on its shape and size. At present, several research studies on CVD reaction using fixed/fluidized bed with horizontal/vertical reactors and CNTs growth parameters are being conducted only in lab scale. Hence, it is of the opinion to further explore the potential of using methane as hydrocarbon source with carbon-based supports for metal catalyst will be an opt route to the bulk and low-cost production of CNTs to meet the future global demand.

ACKNOWLEDGMENTS

The authors gratefully acknowledge the financial support provided by the Ministry of Science, Technology and Innovations (MOSTI) under E-Science Fund (Project A/c no: 6013327) and the Universiti Sains Malaysia (USM) for funding this project under the Research University (RU) Grant Scheme (Project A/c No: 814004) and Student Fellowship.

REFERENCES

1. E. B. Barros, A. Jorio, G. G. Samsonidze et al., "Review on the symmetry-related properties of carbon nanotubes," Physics Reports, vol. 431, no. 6, pp. 261–302, 2006.
2. S. B. Sinnott and R. Andrews, "Carbon nanotubes: synthesis, properties, and applications," Critical Reviews in Solid State and Materials Sciences, vol. 26, no. 3, pp. 145–249, 2001. ·
3. Z. L. Wang, P. Poncharal, and W. A. de Heer, "Measuring physical and mechanical properties of individual carbon nanotubes by in situ TEM," Journal of Physics and Chemistry of Solids, vol. 61, no. 7, pp. 1025–1030, 2000.
4. P. M. Ajayan, L. S. Schadler, C. Giannaris, and A. Rubio, "Single-walled carbon nanotube-polymer composites: strength and weakness," Advanced Materials, vol. 12, no. 10, pp. 750–753, 2000.
5. K. T. Lau, M. Lu, and D. Hui, "Coiled carbon nanotubes: synthesis and their potential applications in advanced composite structures," Composites Part B, vol. 37, no. 6, pp. 437–448, 2006.
6. Z. L. Wang, R. P. Gao, P. Poncharal, W. A. de Heer, Z. R. Dai, and Z. W. Pan, "Mechanical and electrostatic properties of carbon nanotubes and nanowires," Materials Science and Engineering C, vol. 16, no. 1-2, pp. 3–10, 2001.
7. S. Iijima, "Helical microtubules of graphitic carbon," Nature, vol. 354, no. 6348, pp. 56–58, 1991.
8. M. Endo, K. Takeuchi, S. Igarashi, K. Kobori, M. Shiraishi, and H. W. Kroto, "The production and structure of pyrolytic carbon nanotubes (PCNTs)," Journal of Physics and Chemistry of Solids, vol. 54, no. 12, pp. 1841–1848, 1993.
9. T. Guo, P. Nikolaev, A. Thess, D. T. Colbert, and R. E. Smalley, "Catalytic growth of single-walled manotubes by laser vaporization," Chemical Physics Letters, vol. 243, no. 1-2, pp. 49–54, 1995.
10. P. Nikolaev, M. J. Bronikowski, R. K. Bradley et al., "Gas-phase catalytic growth of single-walled carbon nanotubes from carbon monoxide," Chemical Physics Letters, vol. 313, no. 1-2, pp. 91–

97, 1999.

11. J. W. Ward, B. Q. Wei, and P. M. Ajayan, "Substrate effects on the growth of carbon nanotubes by thermal decomposition of methane," Chemical Physics Letters, vol. 376, no. 5-6, pp. 717–725, 2003.
12. B. Zheng, Y. Li, and J. Liu, "CVD synthesis and purification of single-walled carbon nanotubes on aerogel-supported catalyst," Applied Physics A, vol. 74, no. 3, pp. 345–348, 2002.
13. C. E. Baddour and C. Briens, "Carbon nanotube synthesis: a review," International Journal of Chemical Reactor Engineering, vol. 3, 2005.
14. P. Avouris, T. Hertel, R. Martel, T. Schmidt, H. R. Shea, and R. E. Walkup, "Carbon nanotubes: nanomechanics, manipulation, and electronic devices," Applied Surface Science, vol. 141, no. 3-4, pp. 201–209, 1999.
15. K. Tsukagoshi, N. Yoneya, S. Uryu et al., "Carbon nanotube devices for nanoelectronics," Physica B, vol. 323, no. 1–4, pp. 107–114, 2002.
16. G. A. Rivas, M. D. Rubianes, M. C. Rodríguez et al., "Carbon nanotubes for electrochemical biosensing," Talanta, vol. 74, no. 3, pp. 291–307, 2007.
17. Y. Yun, Z. Dong, V. Shanov et al., "Nanotube electrodes and biosensors," Nano Today, vol. 2, no. 6, pp. 30–37, 2007.
18. R. Martel, T. Schmidt, H. R. Shea, T. Hertel, and P. Avouris, "Single- and multi-wall carbon nanotube field-effect transistors," Applied Physics Letters, vol. 73, no. 17, pp. 2447–2449, 1998.
19. J. N. Coleman, U. Khan, W. J. Blau, and Y. K. Gun'ko, "Small but strong: a review of the mechanical properties of carbon nanotube-polymer composites," Carbon, vol. 44, no. 9, pp. 1624–1652, 2006.
20. M. José-Yacamán, M. Miki-Yoshida, L. Rendón, and J. G. Santiesteban, "Catalytic growth of carbon microtubules with fullerene structure," Applied Physics Letters, vol. 62, no. 2, pp. 202–204, 1993.
21. H. Zhu, X. Li, C. Xu, and D. Wu, "Co-synthesis of single-walled

carbon nanotubes and carbon fibers,"Materials Research Bulletin, vol. 37, no. 1, pp. 177–183, 2002.

22. W. Ren, F. Li, J. Chen, S. Bai, and H.-M. Cheng, "Morphology, diameter distribution and Raman scattering measurements of double-walled carbon nanotubes synthesized by catalytic decomposition of methane," Chemical Physics Letters, vol. 359, no. 3-4, pp. 196–202, 2002.

23. Q. Li, H. Yan, J. Zhang, and Z. Liu, "Effect of hydrocarbons precursors on the formation of carbon nanotubes in chemical vapor deposition," Carbon, vol. 42, no. 4, pp. 829–835, 2004.

24. V. K. Varadan and J. Xie, "Large-scale synthesis of multi-walled carbon nanotubes by microwave CVD," Smart Materials and Structures, vol. 11, no. 4, pp. 610–616, 2002.

25. B. Fidalgo, Y. Fernández, L. Zubizarreta et al., "Growth of nanofilaments on carbon-based materials from microwave-assisted decomposition of CH4," Applied Surface Science, vol. 254, no. 11, pp. 3553–3557, 2008.

26. J. H. Yen, I. C. Leu, C. C. Lin, and M. H. Hon, "Synthesis of well-aligned carbon nanotubes by inductively coupled plasma chemical vapor deposition," Applied Physics A, vol. 80, no. 2, pp. 415–421, 2005.

27. T. Ikuno, M. Katayama, N. Yamauchi et al., "Selective growth of straight carbon nanotubes by low-pressure thermal chemical vapor deposition," Japanese Journal of Applied Physics, vol. 43, no. 2, pp. 860–863, 2004.

28. M. Kadlečíková, A. Vojačková, J. Breza, V. Luptáková, M. Michalka, and K. Jesenák, "Bundles of carbon nanotubes grown on sapphire and quartz substrates by catalytic hot filament chemical vapor deposition," Materials Letters, vol. 61, no. 23-24, pp. 4549–4552, 2007.

29. H. Y. Yap, B. Ramaker, A. V. Sumant, and R. W. Carpick, "Growth of mechanically fixed and isolated vertically aligned carbon nanotubes and nanofibers by DC plasma-enhanced hot filament chemical vapor deposition," Diamond & Related Materials, vol. 15, no. 10, pp. 1622–1628, 2006.

30. H. E. Unalan and M. Chhowalla, "Investigation of single-walled

carbon nanotube growth parameters using alcohol catalytic chemical vapour deposition," Nanotechnology, vol. 16, no. 10, pp. 2153–2163, 2005.

31. W. Qian, T. Liu, F. Wei, Z. Wang, and Y. Li, "Enhanced production of carbon nanotubes: combination of catalyst reduction and methane decomposition," Applied Catalysis A, vol. 258, no. 1, pp. 121–124, 2004.

32. T. V. Reshetenko, L. B. Avdeeva, Z. R. Ismagilov, and A. L. Chuvilin, "Catalytic filamentous carbon as supports for nickel catalysts," Carbon, vol. 42, no. 1, pp. 143–148, 2004.

33. R. Bonadiman, M. D. Lima, M. J. de Andrade, and C. P. Bergmann, "Production of single and multi-walled carbon nanotubes using natural gas as a precursor compound," Journal of Materials Science, vol. 41, no. 22, pp. 7288–7295, 2006.

34. K. Y. Tran, B. Heinrichs, J.-F. Colomer, J.-P. Pirard, and S. Lambert, "Carbon nanotubes synthesis by the ethylene chemical catalytic vapour deposition (CCVD) process on Fe, Co, and Fe-Co/Al2O3 sol-gel catalysts," Applied Catalysis A, vol. 318, pp. 63–69, 2007.

35. N. Inami, M. Ambri Mohamed, E. Shikoh, and A. Fujiwara, "Synthesis-condition dependence of carbon nanotube growth by alcohol catalytic chemical vapor deposition method," Science and Technology of Advanced Materials, vol. 8, no. 4, pp. 292–295, 2007.

36. R. Longtin, C. Fauteux, R. Goduguchinta, and J. Pegna, "Synthesis of carbon nanofiber films and nanofiber composite coatings by laser-assisted catalytic chemical vapor deposition," Thin Solid Films, vol. 515, no. 5, pp. 2958–2964, 2007.

37. J.-F. Colomer, J.-M. Benoît, C. Stephan, S. Lefrant, G. Van Tendeloo, and J. B.nagy, "Characterization of single-wall carbon nanotubes produced by CCVD method," Chemical Physics Letters, vol. 345, no. 1-2, pp. 11–17, 2001.

38. J. Kong, A. M. Cassell, and H. Dai, "Chemical vapor deposition of methane for single-walled carbon nanotubes," Chemical Physics Letters, vol. 292, no. 4-6, pp. 567–574, 1998.

39. K. Hernadi, Z. Kónya, A. Siska et al., "The role of zeotype catalyst support in the synthesis of carbon nanotubes by CCVD," Studies

in Surface Science and Catalysis, vol. 142, pp. 541–548, 2002.
40. T. Hiraoka, T. Kawakubo, J. Kimura et al., "Selective synthesis of double-wall carbon nanotubes by CCVD of acetylene using zeolite supports," Chemical Physics Letters, vol. 382, no. 5-6, pp. 679–685, 2003.
41. B. C. Liu, S. C. Lyu, S. I. Jung et al., "Single-walled carbon nanotubes produced by catalytic chemical vapor deposition of acetylene over Fe-Mo/MgO catalyst," Chemical Physics Letters, vol. 383, no. 1-2, pp. 104–108, 2004.
42. C. He, N. Zhao, C. Shi, X. Du, and J. Li, "Carbon nanotubes and onions from methane decomposition using Ni/Al catalysts," Materials Chemistry and Physics, vol. 97, no. 1, pp. 109–115, 2006.
43. N. Zhao, Q. Cui, C. He et al., "Synthesis of carbon nanostructures with different morphologies by CVD of methane," Materials Science and Engineering A, vol. 460-461, pp. 255–260, 2007.
44. Y. Yang, Z. Hu, Y. N. Lü, and Y. Chen, "Growth of carbon nanotubes with metal-loading mesoporous molecular sieves catalysts," Materials Chemistry and Physics, vol. 82, no. 2, pp. 440–443, 2003.
45. K. Kidena, Y. Kamiyama, and M. Nomura, "A possibility of the production of carbon nanotubes from heavy hydrocarbons," Fuel Processing Technology, vol. 89, no. 4, pp. 449–454, 2008.
46. H. Liu, G. Cheng, R. Zheng, Y. Zhao, and C. Liang, "Influence of synthesis process on preparation and properties of Ni/CNT catalyst," Diamond & Related Materials, vol. 15, no. 1, pp. 15–21, 2006.
47. Y. Li, X. Zhang, X. Tao et al., "Growth mechanism of multi-walled carbon nanotubes with or without bundles by catalytic deposition of methane on Mo/MgO," Chemical Physics Letters, vol. 386, no. 1-3, pp. 105–110, 2004.
48. B. C. Liu, S. C. Lyu, T. J. Lee et al., "Synthesis of single- and double-walled carbon nanotubes by catalytic decomposition of methane," Chemical Physics Letters, vol. 373, no. 5-6, pp. 475–479, 2003.
49. R. M. de Almeida, H. V. Fajardo, D. Z. Mezalira et al., "Preparation

and evaluation of porous nickel-alumina spheres as catalyst in the production of hydrogen from decomposition of methane," Journal of Molecular Catalysis A, vol. 259, no. 1-2, pp. 328–335, 2006.

50. R. C. Haddon, J. Sippel, A. G. Rinzler, and F. Papadimitrakopoulos, "Purification and separation of carbon nanotubes," MRS Bulletin, vol. 29, no. 4, pp. 252–241, 2004.

51. Y. Ando, X. Zhao, T. Sugai, and M. Kumar, "Growing carbon nanotubes," Materials Today, vol. 7, no. 9, pp. 22–29, 2004.

52. N. Muradov, "Catalysis of methane decomposition over elemental carbon," Catalysis Communications, vol. 2, no. 3-4, pp. 89–94, 2001.

53. C.-M. Chuang, S. P. Sharma, J.-M. Ting, H.-P. Lin, H. Teng, and C.-W. Huang, "Preparation of sea urchin-like carbons by growing one-dimensional nanocarbon on mesoporous carbons," Diamond & Related Materials, vol. 17, no. 4-5, pp. 606–610, 2008. ·

54. A.-C. Dupuis, "The catalyst in the CCVD of carbon nanotubes-a review," Progress in Materials Science, vol. 50, no. 8, pp. 929–961, 2005.

55. I. Vesselényi, K. Niesz, A. Siska et al., "Production of carbon nanotubes on different metal supported catalysts," Reaction Kinetics and Catalysis Letters, vol. 74, no. 2, pp. 329–336, 2001.·

56. K. Hernadi, A. Fonseca, J. B. Nagy, D. Bernaerts, A. Fudala, and A. A. Lucas, "Catalytic synthesis of carbon nanotubes using zeolite support," Zeolites, vol. 17, no. 5-6, pp. 416–423, 1996.

57. R. Wang, H. Xu, L. Guo, and J. Liang, "Growth of single-walled carbon nanotubes on porous silicon,"Applied Surface Science, vol. 252, no. 20, pp. 7347–7351, 2006.

58. W. Wongwiriyapan, M. Katayama, T. Ikuno et al., "Growth of single-walled carbon nanotubes rooted from Fe/Al nanoparticle array," Japanese Journal of Applied Physics, vol. 44, no. 1, pp. 457–460, 2005.

59. C. He, N. Zhao, C. Shi, X. Du, and J. Li, "Synthesis of binary and triple carbon nanotubes over Ni/Cu/Al2O3 catalyst by chemical vapor deposition," Materials Letters, vol. 61, no. 27, pp. 4940–4943, 2007.

60. W. Z. Li, J. G. Wen, M. Sennett, and Z. F. Ren, "Clean double-walled carbon nanotubes synthesized by CVD," Chemical Physics Letters, vol. 368, no. 3-4, pp. 299–306, 2003.
61. L. Ni, K. Kuroda, L.-P. Zhou et al., "Kinetic study of carbon nanotube synthesis over Mo/Co/MgO catalysts," Carbon, vol. 44, no. 11, pp. 2265–2272, 2006. ·
62. H. Yu, Q. Zhang, Q. Zhang et al., "Effect of the reaction atmosphere on the diameter of single-walled carbon nanotubes produced by chemical vapor deposition," Carbon, vol. 44, no. 9, pp. 1706–1712, 2006.
63. A. Tavasoli, K. Sadagiani, F. Khorashe, A. A. Seifkordi, A. A. Rohani, and A. Nakhaeipour, "Cobalt supported on carbon nanotubes—a promising novel Fischer-Tropsch synthesis catalyst," Fuel Processing Technology, vol. 89, no. 5, pp. 491–498, 2008. ·
64. F. Benissad-Aissani, H. Aït-Amar, M.-C. Schouler, and P. Gadelle, "The role of phosphorus in the growth of vapour-grown carbon fibres obtained by catalytic decomposition of hydrocarbons,"Carbon, vol. 42, no. 11, pp. 2163–2168, 2004.
65. S. H. S. Zein and A. R. Mohamed, "Mn/Ni/TiO2 catalyst for the production of hydrogen and carbon nanotubes from methane decomposition," Energy and Fuels, vol. 18, no. 5, pp. 1336–1345, 2004.
66. M. C. Bahome, L. L. Jewell, K. Padayachy et al., "Fe-Ru small particle bimetallic catalysts supported on carbon nanotubes for use in Fischer-Tröpsch synthesis," Applied Catalysis A, vol. 328, no. 2, pp. 243–251, 2007.
67. L. Barthe, S. Desportes, M. Hemati, K. Philippot, and B. Chaudret, "Synthesis of supported catalysts by dry impregnation in fluidized bed," Chemical Engineering Research and Design, vol. 85, no. 6, pp. 767–777, 2007.
68. H.-A. Ichi-oka, N.-O. Higashi, Y. Yamada, T. Miyake, and T. Suzuki, "Carbon nanotube and nanofiber syntheses by the decomposition of methane on group 8-10 metal-loaded MgO catalysts,"Diamond & Related Materials, vol. 16, no. 4–7, pp. 1121–1125, 2007.
69. S. Takenaka, H. Umebayashi, E. Tanabe, H. Matsune, and M. Kishida, "Specific performance of silica-coated Ni catalysts for

the partial oxidation of methane to synthesis gas," Journal of Catalysis, vol. 245, no. 2, pp. 392–400, 2007.

70. M. C. Bahome, L. L. Jewell, D. Hildebrandt, D. Glasser, and N. J. Coville, "Fischer-Tropsch synthesis over iron catalysts supported on carbon nanotubes," Applied Catalysis A, vol. 287, no. 1, pp. 60–67, 2005.

71. L. Piao, Y. Li, J. Chen, L. Chang, and J. Y. S. Lin, "Methane decomposition to carbon nanotubes and hydrogen on an alumina supported nickel aerogel catalyst," Catalysis Today, vol. 74, no. 1-2, pp. 145–155, 2002.

72. J. M. Xu, X. B. Zhang, Y. Li et al., "Preparation of Mg1−xFexMoO4 catalyst and its application to grow MWNTs with high efficiency," Diamond & Related Materials, vol. 13, no. 10, pp. 1807–1811, 2004.

73. Y. Chen and Y.-S. Lim, "A comparison of different preparation methods of Fe-Mo-Mg-O catalyst for the large-scale synthesis of carbon nanotubes," Materials Science Forum, vol. 510-511, pp. 66–69, 2006.

74. S. P. Turano and J. Ready, "Chemical vapor deposition synthesis of self-aligned carbon nanotube arrays," Journal of Electronic Materials, vol. 35, no. 2, pp. 192–194, 2006. ·

75. W. Qian, T. Liu, F. Wei, Z. Wang, and H. Yu, "Carbon nanotubes containing iron and molybdenum particles as a catalyst for methane decomposition," Carbon, vol. 41, no. 4, pp. 846–848, 2003.

76. Q. He, Q. Lin, L.-Z. Yao, W.-L. Cai, and Q. Zhu, "Effect of growth parameters on morphology of vertically aligned carbon nanotubes," Chinese Journal of Chemical Physics, vol. 20, no. 2, pp. 207–212, 2007.

77. Z. Yu, D. Chen, B. Tøtdal, and A. Holmen, "Effect of catalyst preparation on the carbon nanotube growth rate," Catalysis Today, vol. 100, no. 3-4, pp. 261–267, 2005.

78. Q. Zhang, W. Qian, Q. Wen, Y. Liu, D. Wang, and F. Wei, "The effect of phase separation in Fe/Mg/Al/O catalysts on the synthesis of DWCNTs from methane," Carbon, vol. 45, no. 8, pp. 1645–1650, 2007.

79. R. L. Vander Wal, T. M. Ticich, and V. E. Curtis, "Substrate-support interactions in metal-catalyzed carbon nanofiber growth," Carbon, vol. 39, no. 15, pp. 2277–2289, 2001.
80. J.-F. Colomer, G. Bister, I. Willems et al., "Synthesis of single-wall carbon nanotubes by catalytic decomposition of hydrocarbons," Chemical Communications, no. 14, pp. 1343–1344, 1999.
81. S.-P. Chai, S. H. S. Zein, and A. R. Mohamed, "Synthesizing carbon nanotubes and carbon nanofibers over supported-nickel oxide catalysts via catalytic decomposition of methane," Diamond & Related Materials, vol. 16, no. 8, pp. 1656–1664, 2007.
82. P. Piedigrosso, Z. Konya, J.-F. Colomer, A. Fonseca, G. Van Tendeloo, and J. B. Nagy, "Production of differently shaped multi-wall carbon nanotubes using various cobalt supported catalysts," Physical Chemistry Chemical Physics, vol. 2, no. 1, pp. 163–170, 2000.
83. S.-P. Chai, S. H. S. Zein, and A. R. Mohamed, "Preparation of carbon nanotubes over cobalt-containing catalysts via catalytic decomposition of methane," Chemical Physics Letters, vol. 426, no. 4–6, pp. 345–350, 2006.
84. K. C. Mondal, N. J. Coville, M. J. Witcomb, G. Tejral, and J. Havel, "Boron mediated synthesis of multiwalled carbon nanotubes by chemical vapor deposition," Chemical Physics Letters, vol. 437, no. 1–3, pp. 87–91, 2007.
85. C. Vallés, M. Pérez-Mendoza, P. Castell, M. T. Martínez, W. K. Maser, and A. M. Benito, "Towards helical and Y-shaped carbon nanotubes: the role of sulfur in CVD processes," Nanotechnology, vol. 17, no. 17, pp. 4292–4299, 2006.
86. H. Ago, N. Uehara, N. Yoshihara et al., "Gas analysis of the CVD process for high yield growth of carbon nanotubes over metal-supported catalysts," Carbon, vol. 44, no. 14, pp. 2912–2918, 2006.
87. S.-G. Kang, K.-K. Cho, K.-W. Kim, and G.-B. Cho, "Catalytic growth of single- and double-walled carbon nanotubes from Fe-Mo nanoparticles supported on MgO," Journal of Alloys and Compounds, vol. 449, no. 1-2, pp. 269–273, 2008.
88. L. Ci, H. Zhu, B. Wei, J. Liang, C. Xu, and D. Wu, "Phosphorus—a new element for promoting growth of carbon filaments by the

floating catalyst method," Carbon, vol. 37, no. 10, pp. 1652–1654, 1999.

89. F. Benissad-Aissani, H. Aït-Amar, M.-C. Schouler, and P. Gadelle, "The role of phosphorus in the growth of vapour-grown carbon fibres obtained by catalytic decomposition of hydrocarbons,"Carbon, vol. 42, no. 11, pp. 2163–2168, 2004.

90. S. Zhan, Y. Tian, Y. Cui et al., "Effect of process conditions on the synthesis of carbon nanotubes by catalytic decomposition of methane," China Particuology, vol. 5, no. 3, pp. 213–219, 2007.

91. S. P. Chai, V. M. Sivakumar, S. H. S. Zein, and A. R. Mohamed, "The examination of NiO and CoOxcatalysts supported on Al2O3 and SiO2 for carbon nanotubes production via CCVD of methane,"Carbon Science & Technology, vol. 1, pp. 1–3, 2008.

92. J.-H. Lin, C.-S. Chen, H.-L. Ma, C.-Y. Hsu, and H.-W. Chen, "Synthesis of MWCNTs onCuSO4/Al2O3 using chemical vapor deposition from methane," Carbon, vol. 45, no. 1, pp. 223–225, 2007.

93. F. Ohashi, G. Y. Chen, V. Stolojan, and S. R. P. Silva, "The role of the gas species on the formation of carbon nanotubes during thermal chemical vapour deposition," Nanotechnology, vol. 19, no. 44, 2008.

94. N. Q. Zhao, C. N. He, J. Ding et al., "Bamboo-shaped carbon nanotubes produced by catalytic decomposition of methane over nickel nanoparticles supported on aluminum," Journal of Alloys and Compounds, vol. 428, no. 1-2, pp. 79–83, 2007.

95. N. Zhao, C. He, Z. Jiang, J. Li, and Y. Li, "Physical activation and characterization of multi-walled carbon nanotubes catalytically synthesized from methane," Materials Letters, vol. 61, no. 3, pp. 681–685, 2007.

96. L. K. Noda, N. S. Gonçalves, A. Valentini, L. F. D. Probst, and R. M. de Almeida, "Effect of Ni loading and reaction temperature on the formation of carbon nanotubes from methane catalytic decomposition over Ni/SiO2," Journal of Materials Science, vol. 42, no. 3, pp. 914–922, 2007.

97. I. Bustero, G. Ainara, O. Isabel, M. Roberto, R. Inés, and A. Amaya, "Control of the properties of carbon nanotubes synthesized by CVD for application in electrochemical biosensors," Microchimica Acta, vol. 152, no. 3-4, pp. 239–247, 2006.

Chapter 7

Hydrogen Production Technologies: Current State and Future Developments

Christos M. Kalamaras and Angelos M. Efstathiou

Chemistry Department, University of Cyprus, 1678 Nicosia, Cyprus

ABSTRACT

Hydrogen (H_2) is currently used mainly in the chemical industry for the production of ammonia and methanol. Nevertheless, in the near future, hydrogen is expected to become a significant fuel that will largely contribute to the quality of atmospheric air. Hydrogen as a chemical element (H) is the most widespread one on the earth and as molecular dihydrogen (H_2) can be obtained from a number of sources both renewable and nonrenewable by various processes. Hydrogen global production has so far been dominated by fossil fuels, with the most significant contemporary technologies being the steam reforming of hydrocarbons (e.g., natural gas). Pure hydrogen is also produced by electrolysis of water, an energy demanding process. This work reviews the current technologies used for hydrogen (H_2) production from both fossil and renewable biomass resources, including reforming (steam, partial oxidation, autothermal, plasma, and aqueous phase) and pyrolysis. In addition, other methods for generating hydrogen (e.g., electrolysis of water) and purification methods, such as desulfurization and water-gas shift reactions are discussed.

INTRODUCTION

Hydrogen is the simplest and most abundant element on earth. Hydrogen combines readily with other chemical elements, and it is always found as part of another substance, such as water, hydrocarbon, or alcohol. Hydrogen is also found in natural biomass, which includes plants and animals. For this reason, it is considered as an energy carrier and not as an energy source.

Hydrogen can be produced using diverse, domestic resources, including nuclear, natural gas and coal, biomass, and other renewable sources. The latter include solar, wind, hydroelectric, or geothermal energy. This diversity of domestic energy sources makes hydrogen a promising energy carrier and important for energy security. It is desirable that hydrogen be produced using a variety of resources and process technologies or pathways. The production of hydrogen can be achieved via various process technologies, including thermal (natural gas reforming, renewable liquid and biooil processing, biomass, and coal gasification), electrolytic (water splitting using a variety of energy resources), and photolytic (splitting of water using sunlight through biological and electrochemical materials).

The annual production of hydrogen is estimated to be about 55 million tons with its consumption increasing by approximately 6% per year. Hydrogen can be produced in many ways from a broad spectrum of initial raw materials. Nowadays, hydrogen is mainly produced by the steam reforming of natural gas, a process which leads to massive emissions of greenhouse gases [1, 2]. Close to 50% of the global demand for hydrogen is currently generated via steam reforming of natural gas, about 30% from oil/naphtha reforming from refinery/chemical industrial off-gases, 18% from coal gasification, 3.9% from water electrolysis, and 0.1% from other sources [3]. Electrolytic and plasma processes demonstrate a high efficiency for hydrogen production, but unfortunately they are considered as energy intensive processes [4].

The fundamental question lies in the development of alternative technologies for hydrogen production to those based on fossil fuels, especially for its utilization as a fuel in the transportation sector. This problem can be faced by the utilization of alternative renewable resources and related methods of production, such as the gasification or pyrolysis of biomass, electrolytic, photolytic, and thermal cracking of water. However, it is not possible to consider only the ecological perspective, since for example, photolytic cracking of water is environmentally friendly but its efficiency for industrial use is very low. It is thus clear that the processes to be taken into account must consider not only environmental concerns but also the most favorable economics.

HYDROGEN FROM FOSSIL FUELS

Fossil fuel processing technologies convert hydrogen- containing materials derived from fossil fuels, such as gasoline, hydrocarbons, methanol, or ethanol, into a hydrogen-rich gas stream. Fuel processing of methane (natural gas) is the most common commercial hydrogen production technology today. Most fossil fuels contain a certain amount of sulfur, the removal of which is a significant task in the planning of hydrogen-based economy. As a result, the desulfurization process will also be discussed. In addition, the very promising plasma reforming technology recently developed will also be presented.

Hydrogen gas can be produced from hydrocarbon fuels through three basic technologies: (i) steam reforming (SR), (ii) partial oxidation (POX), and (iii) autothermal reforming (ATR). These technologies produce a great deal of carbon monoxide (CO). Thus, in a subsequent step, one or more chemical reactors are used to largely convert CO into carbon dioxide (CO_2) via the water-gas shift (WGS) and preferential oxidation (PrOx) or methanation reactions, which are described later.

Steam Reforming

Steam reforming is currently one of the most widespread and at the same time least expensive processes for hydrogen production [5]. Its advantage arises from the high efficiency of its operation and the low operational and production costs. The most frequently used raw materials are natural gas and lighter hydrocarbons, methanol, and other oxygenated hydrocarbons [6]. The network of reforming reactions for hydrocarbons and methanol used as feedstock is the following [7]:

$$C_mH_n + mH_2O\,(g) \longrightarrow mCO + (m + 0.5\,n)\,H_2 \quad (1)$$

$$C_mH_n + 2mH_2O\,(g) \longrightarrow mCO_2 + (2m + 0.5\,n)\,H_2 \quad (2)$$

$$CO + H_2O\,(g) \longleftrightarrow CO_2 + H_2 \quad (3)$$

$$CH_3OH + H_2O\,(g) \longleftrightarrow CO_2 + 3H_2 \quad (4)$$

The whole process comprises two stages. In the first stage, the hydrocarbon raw material is mixed with steam and fed in a tubular catalytic reactor [8]. During this process, syngas (H_2/CO gas mixture) is produced with lower content in CO_2 ((1) and (2)). The required reaction temperature is achieved by the addition of oxygen or air for combusting part of the raw material (heating gas) inside the reactor. In the second stage, the cooled product gas is fed into the CO catalytic converter, where carbon monoxide is converted to a large extent by means of steam into carbon dioxide and hydrogen (3). The steam reforming catalytic process requires a raw material free of sulfur-containing compounds in order to avoid deactivation of the catalyst used.

The SR process requires modest temperatures, for example, 180°C for methanol and oxygenated hydrocarbons and more than 500°C for most conventional hydrocarbons [9, 10]. The catalysts used can be divided into two types: nonprecious metal (typically nickel) and precious metals from Group VIII elements (typically platinum or rhodium). Due to severe mass and heat transfer limitations, conventional steam reformers are limited by the effectiveness factor of pelletized catalysts, which is typically less than 5% [11]. Therefore, kinetics is rarely the limiting factor with conventional steam reformer reactors [12], and, therefore, less expensive nickel catalysts are used industrially.

An important factor characterizing the SR process is the H:C atom ratio in the feedstock material. The higher this ratio is the lower carbon dioxide emission is formed. A membrane reactor can replace both reactors in a conventional SR process for achieving the overall reaction (2) [13]. The heat efficiency of hydrogen production by the SR of methane process on an industrial scale is around 70–85% [14]. A number of other raw materials are also possible to achieve this efficiency in the near future, such as solid communal waste, wastes from food industry, oils, purposefully cultivated or waste agricultural biomass, and fuels of fossil origin such as coal. The disadvantage is the high production of CO_2, ca. 7.05 kg CO_2/kg H_2.

Partial Oxidation

Partial oxidation (POX) and catalytic partial oxidation (CPOX) of hydrocarbons have been proposed in hydrogen production for automobile fuel cells and some other commercial applications [15, 16]. The gasified raw material can be methane and biogas but primarily heavy oil fractions (e.g., vacuum remnants, heating oil), whose further treatment and utilization are difficult [17]. POX is a noncatalytic process, in which the raw material is gasified in the presence of oxygen ((5) and (6)) and possibly steam ((7), ATR) at temperatures in the 1300–1500°C range and pressures in the 3–8 MPa range. In comparison with the steam reforming (H_2 : CO = 3 : 1), more CO is produced (H_2 : CO = 1 : 1 or 2 : 1). The process is therefore complemented by the conversion of CO with steam into H_2 and CO_2. This reaction contributes to the maintenance of equilibrium between the individual reaction products [18]:

$$CH_4 + O_2 \longrightarrow CO + 2H_2 \tag{5}$$

$$CH_4 + 2O_2 \longrightarrow CO_2 + 2H_2O \tag{6}$$

$$CH_4 + H_2O\,(g) \longrightarrow CO + 3H_2 \tag{7}$$

The gaseous mixture formed through partial oxidation contains CO, CO_2, H_2O, H_2, CH_4, hydrogen sulfide (H_2S), and carbon oxysulfide (COS). A part of the gas is burned to provide enough heat for the endothermic processes. The soot created by the decomposition of acetylene as an intermediate product is an undesired product. Its amount depends on the proportion of H:C in the initial raw fuel material. There has been, therefore, like with the SR, an endeavor to shift to raw materials containing a higher H:C ratio, for example, to natural gas. While the operation of the reactor is less expensive in comparison with the steam reforming, the subsequent conversion makes this technology more expensive. Since the process does not require the use of a catalyst, it is not necessary to remove sulfurous elements from natural gas, which lowers the efficiency of the catalyst. The sulfurous compounds contained in the gasified raw material are converted into hydrogen sulfide (ca. 95%) and carbon oxysulfide (ca. 5%) [19].

Catalysts can be added to the partial oxidation system (CPOX) in order to lower the operating temperature, ca. 700–1000°C. However, temperature control is proving hard because of coke and hot spot formation due to the exothermic nature of the reactions [10, 15, 16, 20]. For natural gas conversion, the catalysts are typically based on Ni or Rh. However, nickel has a strong tendency to coke, and the cost of Rh has increased significantly. Krummenacher et al. [16] had succeeded in using catalytic partial oxidation for decane, hexadecane, and diesel fuel. The high operating temperatures (>800°C) [16] and safety concerns may make their use for practical and compact portable devices difficult due to thermal management [21]. Typically, the thermal efficiency of POX reactors with methane as fuel lies in the range of 60–75% [22].

Autothermal Reforming

As previously mentioned, in the autothermal reforming (ATR), steam is added in the catalytic partial oxidation process. ATR is a combination of both steam reforming (endothermic) and partial oxidation (exothermic) reactions [23]. ATR has the advantages of not requiring external heat and being simpler and less expensive than SR of methane.

The range of operation of a fuel processor for hydrogen production is depicted in Figure 1. The selection of operation conditions of the reformer depends on the specific target. A main target is the high hydrogen yield with low carbon monoxide content. Maximum hydrogen efficiency and low carbon monoxide content are possible for steam reforming. However, steam reforming is an endothermic process and therefore energy demanding. This energy has to be transferred into the system from the outside.

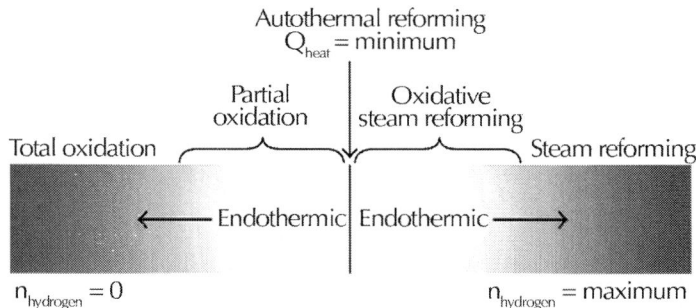

Figure 1: Operating conditions for POX, ATR, and SR.

Another significant advantage of ATR over SR process is that it can be shut down and started very rapidly, while producing a larger amount of hydrogen than POX alone [23]. There are some expectations that this process will become attractive for the "Gas to Liquid" fuel industry due to favorable gas composition for the Fischer-Tropsch synthesis, ATR's relative compactness, lower capital cost, and the potential for economies of scale [24]. For methane reforming, the thermal efficiency is comparable to that of POX (ca. 60–75%) and slightly less than that of steam reforming. Gasoline and other higher hydrocarbons may be converted into hydrogen on board for use in automobiles by the autothermal process, using suitable catalysts [25].

Water-Gas Shift, Preferential Oxidation, and Methanation

The reforming process produces a product gas mixture with significant concentrations of carbon monoxide, often 5 vol% or more (ca. 10 vol%) [10]. To increase the amount of hydrogen, the product gas is passed through a water-gas shift (WGS) reactor to decrease the carbon monoxide content, while at the same time increasing the hydrogen content (3). Typically, a high temperature is desired in order to favor fast kinetics. However, this results in high equilibrium carbon monoxide selectivity and decreased hydrogen product yield. Therefore, the reduction in the

CO content of syngas is achieved in a two-step process that involves a high- and a low-temperature water-gas shift reaction, known as "HTS" and "LTS" processes, respectively (Figure 2). In the first step, carried out in the 310–450°C range with the use of Fe_3O_4/Cr_2O_3 catalyst, the CO concentration is reduced from 10 to 3 vol%. In the second step, carried out in the 180–250°C range, the CO content is further reduced to the low level of 500 ppm using $Cu/ZnO/Al_2O_3$ catalysts [26].

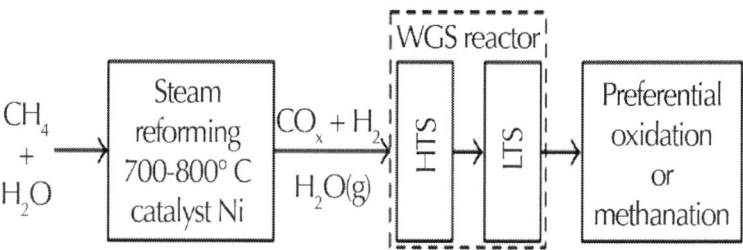

Figure 2: A diagram of methane steam reforming with subsequent carbon monoxide conversion into carbon dioxide and hydrogen.

To further reduce the carbon monoxide content in the product gas, a preferential oxidation (PrOx) reactor or a carbon monoxide selective methanation reactor is used [10, 27]. Sometimes, the term selective oxidation is used in place of preferential oxidation. Selective oxidation refers to carbon monoxide reduction within a fuel cell, typically a proton exchange membrane (PEM) fuel cell, whereas preferential oxidation occurs in a reactor external to the fuel cell [27]. The PrOx and methanation reactors have their own advantages and challenges. The preferential oxidation reactor increases the system's complexity because precise concentrations of air must be added to the system [10, 27]. However, these reactors are compact, and if excessive air is introduced, some hydrogen is burned.

Methanation reactors are simpler in that no air is required. However, for every CO reacted, three H_2 molecules are consumed. Also, CO_2 reacts with hydrogen, and careful control of reactor's conditions needs to be maintained in order to minimize unnecessary consumption of hydrogen. Currently, preferential oxidation is the primary technique in development [27]. The catalysts are typically noble metals such as platinum, ruthenium, or rhodium supported on Al_2O_3 [10, 27]. At the same time, H_2 is purified by alternative approaches, namely pressure-swing adsorption, cryogenic distillation, and membrane technologies, which can ensure the necessary purity of hydrogen (ca. 98-99%). The most advantageous gas-purification method is the pressure-swing adsorption for its high efficiency (>99.99%) and flexibility.

Desulfurization

As discussed before, current hydrogen production arises primarily from the processing of natural gas, although with the substantial advances in fuel cells, there is an increased attention to other fuels, such as methanol, propane, gasoline, and logistic fuels, such as jet-A, diesel, and JP8 [28]. With the exception of methanol, all of these fuels contain some amount of sulfur, with the specific sulfur-containing compounds dependent on the fuel type and source. For this reason, desulfurization is considered as a very important step in fuel processing technologies.

The desulfurization processes can be classified based on the nature of the key physicochemical process used for sulfur removal (Figure 3). The most developed and commercialized technologies are those that catalytically convert organosulfur compounds with sulfur elimination. Such catalytic conversion technologies include conventional hydrodesulfurization (HDS), hydrotreating with advanced catalysts and/or reactor designs, and a combination of hydrotreating with some additional chemical processes to maintain fuel specifications [29, 30]. The main feature of the technologies of the second type is the application of physicochemical processes different in nature from catalytic HDS to separate and/or transform organosulfur compounds from refinery streams. Such technologies include, as a key step, distillation, alkylation, oxidation, extraction, adsorption, or combination of these [31].

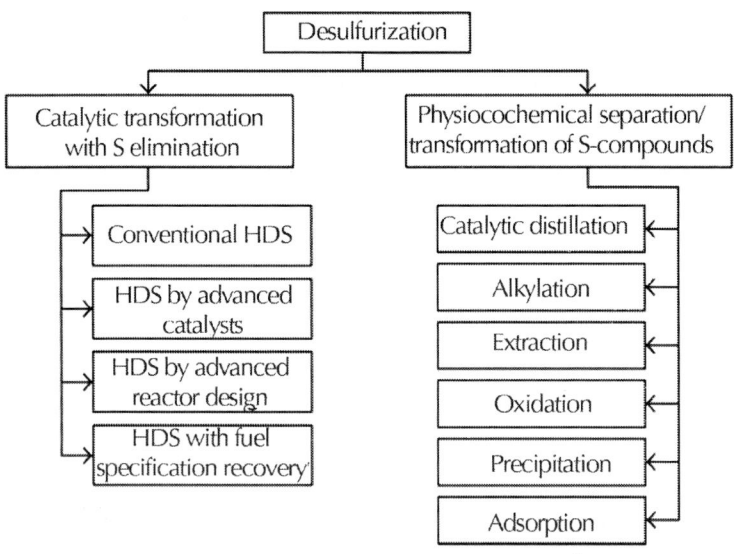

Figure 3: Desulfurization technologies classified by the nature of the key process to remove sulfur.

Plasma Reforming

In the case of plasma reforming, the network of reforming reactions is the same as that in conventional reforming. However, energy and free radicals used for the reforming reaction are provided by plasma typically generated with electricity or heat [32–35]. When water or steam is injected with the fuel, H, OH, and O radicals in addition to electrons are formed, thus creating conditions for both reductive and oxidative reactions to occur. Plasma reforming technologies have been developed to facilitate POX, ATR, and steam reforming, with the majority of the reactors being POX and ATR [35]. There are essentially two main categories of plasma reforming, namely, thermal and nonthermal [35].

Plasma devices referred to as plasmatrons can generate very high temperatures (ca. >2000°C) with a high degree of control using electricity [32–35]. The heat generated is independent of reaction chemistry, and optimal operating conditions can be maintained over a wide range of feed rates and gas compositions. Compactness of the plasma reformer is ensured by high energy density associated with the plasma itself, and by the reduced reaction times, resulting in short residence times. Hydrogen-rich gas streams can be efficiently produced in plasma reformers from a variety of hydrocarbon fuels (e.g., gasoline, diesel, oil, biomass, natural gas, and jet fuel) with conversion efficiencies close to 100% [32, 36]. The plasma reforming technology has potential advantages over conventional technologies of hydrogen manufacturing [32–35]. The plasma conditions (e.g., high temperatures, high degree of dissociation, and substantial degree of ionization) can be used to accelerate thermodynamically favorable chemical reactions without a catalyst or provide the energy required for endothermic reforming processes to occur. Plasma reformers can provide a number of advantages, namely compactness and low weight (due to high power density), high conversion efficiencies, minimal cost (simple metallic or carbon electrodes and simple power supplies), fast response time (fraction of a second), operation with a broad range of fuels, including heavy hydrocarbons (crude) and "dirty" hydrocarbons (high sulfur diesel). This technology could be used to manufacture hydrogen for a variety of stationary applications, such as distributed and low-pollution electricity generation for fuel cells [32]. It could also be used for mobile applications (e.g., on-board generation of hydrogen for fuel cell powered vehicles) and for refueling applications (e.g., stationary sources of hydrogen for vehicles).

The only disadvantages of plasma reforming are the dependence on electricity and the difficulty of high-pressure operation (required for high-pressure processes such as ammonia production). High pressure, while achievable, increases electrode erosion due to decreased arc mobility and, therefore, it decreases electrode lifetime [33].

HYDROGEN FROM RENEWABLE SOURCES

Hydrogen could be also produced by other methods than reforming of fossil fuels. A brief description of the biomass-based approaches (e.g., gasification, pyrolysis, and aqueous phase reforming) along with production of hydrogen from water (e.g., electrolysis, photoelectrolysis, and thermochemical water splitting) is described later.

Biomass Gasification

In the near term, biomass is anticipated to become the most likely renewable organic substitute to petroleum. Biomass is available from a wide range of sources, such as animal wastes, municipal solid wastes, crop residues, short rotation woody crops, agricultural wastes, sawdust, aquatic plants, short rotation herbaceous species (e.g., switch grass), waste paper, corn, and many others [37, 38].

Gasification technology commonly used with biomass and coal as fuel feedstock is very mature and commercially used in many processes. It is a variation of pyrolysis, and, therefore, is based upon partial oxidation of the feedstock material into a mixture of hydrogen, methane, higher hydrocarbons, carbon monoxide, carbon dioxide, and nitrogen, known as "producer gas" [37]. The gasification process typically suffers from low thermal efficiency since moisture contained in the biomass must also be vaporized. It can be performed with or without a catalyst and in a fixed-bed or fluidized-bed reactor, with the latter reactor having typically better performance [38]. Addition of steam and/or oxygen in the gasification process results in the production of "syngas" with a H_2/CO ratio of 2/1, the latter used as feedstock to a Fischer-Tropsch reactor to make higher hydrocarbons (synthetic gasoline and diesel) or to a WGS reactor for hydrogen production [38]. Superheated steam (ca. 900°C) has been used to reform dry biomass to achieve high hydrogen yields. However, gasification process provides significant amounts of "tars" (a complex mixture of higher aromatic hydrocarbons) in the product gas even operated in the 800–1000°C range. A secondary reactor, which utilizes calcined dolomite and/or nickel catalyst, is used to catalytically clean and upgrade the product gas [38]. Ideally, oxygen should be used in these gasification plants; however, oxygen separation unit is cost prohibitive for small-scale plants. This limits the gasifiers to the use of air resulting in significant dilution of the product as well as the production of NO_x. Low-cost, efficient oxygen separators are needed for this technology. For hydrogen production, a WGS process can be employed to increase the hydrogen concentration followed by a separation process to produce pure hydrogen [39]. Typically, gasification reactors are built on a large scale and require massive amounts of material to be continuously fed. They can achieve efficiencies in the order of 35–50% based on the lower heating value [4]. One of the problems of this technology is that a tremendous amount of resources must be used to gather the large amounts of biomass to the central processing plant. Currently, the high logistics costs of gasification plants and the removal of "tars" to acceptable levels for pure hydrogen production limit the commercialization of biomass-based hydrogen production. Future development of smaller efficient distributed gasification plants may be required for this technology for cost effective hydrogen production.

Pyrolysis and Copyrolysis

Another currently promising method of hydrogen production is pyrolysis or copyrolysis. Raw organic material is heated and gasified at a pressure of 0.1–0.5 MPa in the 500–900°C range [40–43]. The process takes place in the absence of oxygen and air, and therefore the formation of dioxins can be almost ruled out. Since no water or air is present, no carbon oxides (e.g., CO or CO_2) are formed, eliminating the need for secondary reactors (WGS, PrOx, etc.). Consequently, this process offers significant emissions reduction. However, if air or water is present (the materials have not been dried), significant CO_x emissions will be produced. Among the advantages of this process are fuel flexibility, relative simplicity and compactness, clean carbon byproduct, and reduction in CO_x emissions [40–43]. The reaction can be generally described by the following equation: [41]

$$C_nH_m + \text{heat} \longrightarrow nC + 0.5\, m\, H_2 \qquad (8)$$

Based on the temperature range, pyrolysis processes are divided into low (up to 500°C), medium (500–800°C), and high temperatures (over 800°C). Fast pyrolysis is one of the latest processes for the transformation of organic material into products with higher energy content. The products of fast pyrolysis appear in the entire phases formed (solid, liquid, and gaseous). One of the challenges with this approach is the potential for fouling by the carbon formed, but proponents claim that this can be minimized by appropriate design. Since it has the potential for lower CO and CO_2 emissions, and it can be operated in such a way as to recover a significant amount of solid carbon, which is easily sequestered [41, 44], pyrolysis may play a significant role in the future. The application of the copyrolysis of a mixture of coal with organic wastes has recently received an interest in industrially advanced countries, as it should limit and lighten the burden of wastes in waste disposal (waste and pure plastics, rubber, cellulose, paper, textiles, and wood) [45, 46]. Pyrolysis and copyrolysis are well-developed processes and could be used in commercial scale.

Aqueous Phase Reforming

Aqueous phase reforming (APR) is a technology under development to process oxygenated hydrocarbons or carbohydrates of renewable biomass resources to produce hydrogen [47, 48], as depicted in Figure 4. The APR reactions take place at substantially lower temperatures (220–270°C) than conventional alkane steam reforming (ca. 600°C). The low temperatures at which aqueous-phase reforming reactions occur minimize undesirable decomposition reactions typically encountered when carbohydrates are heated to elevated temperatures [49, 50]. Also, the water-gas shift reaction (WGS) is favorable at the same temperatures as in APR reactions, thus making it possible to generate H_2 and CO_2 in a single reactor with low amounts of CO. In contrast, typical steam reforming processes require multistage or multiple reactors to achieve low levels of CO in the product gas. Another advantage of the APR process is that it eliminates the need to vaporize water, which represents a major energy saving compared to conventional, vapor-phase steam reforming processes. Most of the research to date has been focused on supported Group VIII catalysts, with Pt-containing solids having the highest catalytic activity. Even though they have lower activity, nickel-based catalysts have been evaluated due to nickel's low cost [47]. The advocates of this technology claim that this technology is more amiable to efficiently and selectively converting biomass feedstock to hydrogen. Aqueous feed concentrations of 10–60 wt% were reported for glucose and glycols [51]. Catalyst selection is important to avoid methanation, which is thermodynamically favorable, along with Fischer-Tropsch products, such as propane, butane, and hexane [48, 52]. Recently, Rozmiarek [53] reported an aqueous phase reformer-based process that achieved an efficiency larger than 55% with a feed composed of 60 wt% glucose in water. However, the catalyst was not stable during long-term testing (200 days on stream) [53]. Finally, due to moderate space time yields, these reactors tend to be somewhat large. Improving catalyst activity and durability is an area where significant progress can be made.

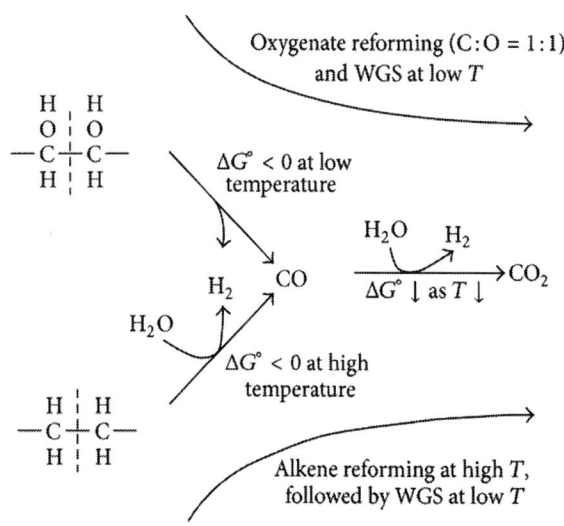

Figure 4: Reaction pathways for the production of H_2 by reactions of oxygenated hydrocarbons with water during APR.

Electrolysis

A promising method for the production of hydrogen in the future could be water electrolysis. Currently, approximately only 4% of hydrogen worldwide is produced by this process [2]. The electrolysis of water or its breaking into hydrogen and oxygen is a well-known method which began to be used commercially already in 1890.

Electrolysis is a process in which a direct current passing through two electrodes in a water solution results in the breaking of the chemical bonds present in water molecule into hydrogen and oxygen:

$$\text{Cathode}: 2H_2O\,(l) + 2e^- \longrightarrow H_2\,(g) + 2OH^-\,(aq) \tag{9}$$

$$\text{Anode}: 4OH^-\,(aq) \longrightarrow O_2\,(g) + 2H_2O\,(l) + 4e^- \tag{10}$$

$$\text{Overall}: 2H_2O \longrightarrow 2H_2 + O_2 \tag{11}$$

The electrolysis process takes place at room temperature. A commonly used electrolyte in water electrolysis is sulfuric acid, and the electrodes are of platinum (Pt), which does not react with sulfuric acid. The process is ecologically clean because no greenhouse gases are formed, and the oxygen produced has further industrial applications. However, in comparison with the foregoing methods described, electrolysis is a highly energy-demanding technology.

The energetic efficiency of the electrolysis of water (chemical energy acquired per electrical energy supplied) in practice reaches 50–70% [54]. It is essentially the conversion of electrical energy to chemical energy in the form of hydrogen, with oxygen as a useful byproduct. The most common electrolysis technology is alkaline-based but proton exchange membrane (PEM) and solid oxide electrolysis cells (SOEC) have been developed [55, 56]. SOEC electrolyzers are the most electrically efficient but the least developed. SOEC technology has challenges with corrosion, seals, thermal cycling, and chrome migration. PEM electrolyzers are more efficient than alkaline and do not have corrosion problems and seals issues as SOEC; however, they cost more than alkaline systems. Alkaline systems are the most developed and the lowest in capital cost. They have the lowest efficiency, so they have the highest electrical energy cost.

Photoelectrolysis

Photoelectrolysis is one of the renewable ways of hydrogen production, exhibiting promising efficiency and costs, although it is still in the phase of experimental development [57]. Currently, it is the least expensive and the most effective method of hydrogen production from renewable resources. The photoelectrode is a semiconducting device absorbing solar energy and simultaneously creating the necessary voltage for the direct decomposition of water molecule into oxygen and hydrogen. Photoelectrolysis utilizes a photoelectrochemical (PEC) light collection system for driving the electrolysis of water. If the semiconductor photoelectrode is submerged in an aqueous electrolyte exposed to solar radiation, it will generate enough electrical energy to support the generated reactions of hydrogen and oxygen. When generating hydrogen, electrons are released into the electrolyte, whereas the generation of oxygen requires free electrons. The reaction depends on the type of semiconductor material and on the solar intensity, which produces a current density of 10–30 mA/cm². At these current densities, the voltage necessary for electrolysis is approximately 1.35 V.

The photoelectrode is comprised of photovoltaic (semiconductor), catalytic and protective layers, which can be modeled as independent components [58]. Each layer influences the overall efficiency of the photoelectrochemical system. The photovoltaic layer is produced from light absorbing semiconductor materials. The light absorption of the semiconductor material is directly proportional to the performance of the photoelectrode. Semiconductors with wide bands provide the necessary potential for the splitting of water [54].

The catalytic layers of the photoelectrochemical cell also influence the performance of the electrolysis and require suitable catalysts for water splitting. The encased layer is another important component of the photoelectrode which prevents the semiconductor from corroding inside the aqueous electrolyte. This layer must be highly transparent in order to be able to provide the maximum solar energy, so that it could reach the photovoltaic semiconducting layer.

Thermochemical Water Splitting

Thermochemical cycles have been developed already since the 1970s and 1980s when they had to contribute to the search for new sources of production of alternative fuels during the petroleum crisis. In thermochemical water splitting, also called thermolysis, heat alone is used to decompose water to hydrogen and oxygen [59]. It is believed that overall efficiencies close to 50% can be achieved using these processes [60].

The single-step thermal dissociation of water is described as follows:

One drawback of this process comes from the need of an effective technique to separate H_2 and O_2 to avoid ending up with an explosive mixture. Semipermeable membranes based on ZrO_2 and other high-temperature materials can be used for this purpose. Separation can also be achieved after the product gas mixture is quenched to lower temperatures. Palladium membranes can then be used for effective hydrogen separation.

It is well known that water will decompose at 2500°C, but materials stable at this temperature and also sustainable heat sources are not easily available. Therefore, chemical reagents have been proposed to lower the temperature, whereas more than 300 water-splitting cycles were referenced in the literature [61]. All of the processes have significantly reduced the operating temperature to lower than 2500°C, but typically require higher pressures. However, it is believed that scaling-up the processes may lead to improved thermal efficiency, overcoming one of the main challenges faced by this technology. In addition, a better understanding of the relationship between capital cost, thermodynamic losses, and process thermal efficiency may lead to reduced hydrogen production costs [60].

ECONOMIC ASPECTS ON HYDROGEN PRODUCTION

At present, the most widely used and cheapest method for hydrogen production is the steam reforming of methane (natural gas). This method includes about half of the world hydrogen production, and hydrogen price is about 7 USD/GJ. A comparable price for hydrogen is provided by partial oxidation of hydrocarbons. However, greenhouse gases generated by thermochemical processes must be captured and stored, and thus, an increase in the hydrogen price by 25–30% must be considered [62].

The further used thermochemical processes include gasification and pyrolysis of biomass. The price of hydrogen thus obtained is about three times greater than the price of hydrogen obtained by the SR process. Therefore, these processes are generally not considered as cost competitive of steam reforming. The price of hydrogen from gasification of biomass ranges from 10–14 USD/GJ and that from pyrolysis 8.9–15.5 USD/GJ. It depends on the equipment, availability, and cost of feedstock [1].

Electrolysis of water is one of the simplest technologies for producing hydrogen without byproducts. Electrolytic processes can be classified as highly effective. On the other hand, the input electricity cost is relatively high and plays a key role in the price of hydrogen obtained.

By the year 2030, the dominant methods for hydrogen production will be steam reforming of natural gas and catalyzed biomass gasification.

In a relatively small extent both coal gasification and electrolysis will be used. The use of solar energy in a given context is questionable but also possible. Probably, the role of solar energy will increase by 2050 [1].

CONCLUSIONS

There is a tremendous amount of research being pursued towards the development of hydrogen (H_2) generation technologies. Currently, the most developed and used technology is the reforming of hydrocarbons. In order to decrease the dependence on fossil fuels, significant developments in other H_2 generation technologies from renewable resources such as biomass and water are considered. Table 1 summarizes the technologies along with their feedstock used and efficiencies obtained. It is important to note that H_2 can be produced from a wide variety of feedstock available almost everywhere. There are many processes under development with minimal environmental impact. Development of these technologies may decrease the world's dependence on fuels that come primarily from unstable regions. The "in-house" H_2 production may increase both national energy and economic security. The ability of H_2 to be produced from a wide variety of feedstock and using a wide variety of processes may make every region of the world able to produce much of its own energy. It is clear that as the technologies develop and mature, H_2 may prove to be the most ubiquitous fuel available.

Table 1: Hydrogen production technologies summary

Technology	Feedstock	Efficiency	Maturity
Steam reforming	Hydrocarbons	70–85%	Commercial
Partial oxidation	Hydrocarbons	60–75%	Commercial
Autothermal reforming	Hydrocarbons	60–75%	Near term
Plasma reforming	Hydrocarbons	9–85%*	Long term
Biomass gasification	Biomass	35–50%	Commercial
Aqueous phase reforming	Carbohydrates	35–55%	Med. term
Electrolysis	H_2O + electricity	50–70%	Commercial
Photolysis	H_2O + sunlight	0.5%*	Long term
Thermochemical water splitting	H_2O + heat	NA	Long term

*Hydrogen purification is not included.

ACKNOWLEDGMENTS

The authors gratefully acknowledge the Research Committee of the University of Cyprus and the Cyprus Research Promotion Foundation for their financial support in developing fundamental research on catalytic hydrogen production technologies.

REFERENCES

1. M. Balat and M. Balat, "Political, economic and environmental impacts of biomass-based hydrogen,"International Journal of Hydrogen Energy, vol. 34, no. 9, pp. 3589–3603, 2009.
2. A. Konieczny, K. Mondal, T. Wiltowski, and P. Dydo, "Catalyst development for thermocatalytic decomposition of methane to hydrogen," International Journal of Hydrogen Energy, vol. 33, no. 1, pp. 264–272, 2008.
3. N. Z. Muradov and T. N. Veziro lu, "From hydrocarbon to hydrogen-carbon to hydrogen economy,"International Journal of Hydrogen Energy, vol. 30, no. 3, pp. 225–237, 2005.
4. J. D. Holladay, J. Hu, D. L. King, and Y. Wang, "An overview of hydrogen production technologies,"Catalysis Today, vol. 139, no. 4, pp. 244–260, 2009.
5. J. M. Ogden, M. M. Steinbugler, and T. G. Kreutz, "Comparison of hydrogen, methanol and gasoline as fuels for fuel cell vehicles: implications for vehicle design and infrastructure development," Journal of Power Sources, vol. 79, no. 2, pp. 143–168, 1999.
6. M. Onozaki, K. Watanabe, T. Hashimoto, H. Saegusa, and Y. Katayama, "Hydrogen production by the partial oxidation and steam reforming of tar from hot coke oven gas," Fuel, vol. 85, no. 2, pp. 143–149, 2006.
7. J. R. Rostrup-Nielsen, "Conversion of hydrocarbons and alcohols for fuel cells," Physical Chemistry Chemical Physics, vol. 3, no. 3, pp. 283–288, 2001.
8. H. Song, L. Zhang, R. B. Watson, D. Braden, and U. S. Ozkan, "Investigation of bio-ethanol steam reforming over cobalt-based catalysts," Catalysis Today, vol. 129, no. 3-4, pp. 346–354, 2007.
9. R. Farrauto, S. Hwang, L. Shore et al., "New material needs for

hydrocarbon fuel processing: generating hydrogen for the PEM fuel cell," Annual Review of Materials Research, vol. 33, pp. 1–27, 2003.

10. C. Song, "Fuel processing for low-temperature and high-temperature fuel cells: challenges, and opportunities for sustainable development in the 21st century," Catalysis Today, vol. 77, no. 1-2, pp. 17–49, 2002.

11. A. M. Adris, B. B. Pruden, C. J. Lim, and J. R. Grace, "On the reported attempts to radically improve the performance of the steam methane reforming reactor," Canadian Journal of Chemical Engineering, vol. 74, no. 2, pp. 177–186, 1996.

12. J. Rostrup-Nielsen, "Hydrogen generation by catalysis," in Encyclopedia of Catalysis, I. T. Horvath, Ed., Wiley Interscience, 2003.

13. Y. Shirasaki, T. Tsuneki, Y. Ota et al., "Development of membrane reformer system for highly efficient hydrogen production from natural gas," International Journal of Hydrogen Energy, vol. 34, no. 10, pp. 4482–4487, 2009.

14. B. Sorensen, Hydrogen and Fuel Cells, Academic Press, 2011.

15. K. L. Hohn and L. D. Schmidt, "Partial oxidation of methane to syngas at high space velocities over Rh-coated spheres," Applied Catalysis A, vol. 211, no. 1, pp. 53–68, 2001.

16. J. J. Krummenacher, K. N. West, and L. D. Schmidt, "Catalytic partial oxidation of higher hydrocarbons at millisecond contact times: decane, hexadecane, and diesel fuel," Journal of Catalysis, vol. 215, no. 2, pp. 332–343, 2003.

17. A. Holmen, "Direct conversion of methane to fuels and chemicals," Catalysis Today, vol. 142, no. 1-2, pp. 2–8, 2009.

18. K. Aasberg-Petersen, J. H. Bak Hansen, T. S. Christensen et al., "Technologies for large-scale gas conversion," Applied Catalysis A, vol. 221, no. 1-2, pp. 379–387, 2001.

19. A. E. Lutz, R. W. Bradshaw, L. Bromberg, and A. Rabinovich, "Thermodynamic analysis of hydrogen production by partial oxidation reforming," International Journal of Hydrogen Energy, vol. 29, no. 8, pp. 809–816, 2004.

20. L. Pino, V. Recupero, S. Beninati, A. K. Shukla, M. S. Hegde,

and P. Bera, "Catalytic partial-oxidation of methane on a ceria-supported platinum catalyst for application in fuel cell electric vehicles," Applied Catalysis A, vol. 225, no. 1-2, pp. 63–75, 2002.
21. J. D. Holladay, Y. Wang, and E. Jones, "Review of developments in portable hydrogen production using microreactor technology," Chemical Reviews, vol. 104, no. 10, pp. 4767–4790, 2004.
22. T. A. Semelsberger, L. F. Brown, R. L. Borup, and M. A. Inbody, "Equilibrium products from autothermal processes for generating hydrogen-rich fuel-cell feeds," International Journal of Hydrogen Energy, vol. 29, no. 10, pp. 1047–1064, 2004.
23. F. Joensen and J. R. Rostrup-Nielsen, "Conversion of hydrocarbons and alcohols for fuel cells," Journal of Power Sources, vol. 105, no. 2, pp. 195–201, 2002.
24. D. J. Wilhelm, D. R. Simbeck, A. D. Karp, and R. L. Dickenson, "Syngas production for gas-to-liquids applications: technologies, issues and outlook," Fuel Processing Technology, vol. 71, no. 1–3, pp. 139–148, 2001.
25. S. Ayabe, H. Omoto, T. Utaka et al., "Catalytic autothermal reforming of methane and propane over supported metal catalysts," Applied Catalysis A, vol. 241, no. 1-2, pp. 261–269, 2003.
26. C. Rhodes, B. P. Williams, F. King, and G. J. Hutchings, "Promotion of Fe_3O_4/Cr_2O_3 high temperature water gas shift catalyst," Catalysis Communications, vol. 3, no. 8, pp. 381–384, 2002.
27. P. Pietrogrande and M. Bezzeccheri, "Fuel processing," in Fuel Cell Systems, L. J. M. J. Blomen and M. N. Mugerwa, Eds., pp. 121–156, Plenum Press, New York, NY, USA, 1993.
28. M. W. Iwigg, Catalyst Handbook, Wolfe Publishing, London, UK, 1989.
29. I. V. Babich and J. A. Moulijn, "Science and technology of novel processes for deep desulfurization of oil refinery streams: a review," Fuel, vol. 82, no. 6, pp. 607–631, 2003.
30. C. Song, "An overview of new approaches to deep desulfurization for ultra-clean gasoline, diesel fuel and jet fuel," Catalysis Today, vol. 86, no. 1–4, pp. 211–263, 2003.
31. A. J. Hernández-Maldonado and R. T. Yang, "Desulfurization of

liquid fuels by adsorption via p complexation with Cu(I)-Y and Ag-Y zeolites," Industrial & Engineering Chemistry Research, vol. 42, no. 1, pp. 123–129, 2002.

32. L. Bromberg, D. R. Cohn, and A. Rabinovich, "Plasma reformer-fuel cell system for decentralized power applications," International Journal of Hydrogen Energy, vol. 22, no. 1, pp. 83–94, 1997.

33. L. Bromberg, D. R. Cohn, A. Rabinovich, and N. Alexeev, "Plasma catalytic reforming of methane," International Journal of Hydrogen Energy, vol. 24, no. 12, pp. 1131–1137, 1999.

34. T. Hammer, T. Kappes, and M. Baldauf, "Plasma catalytic hybrid processes: gas discharge initiation and plasma activation of catalytic processes," Catalysis Today, vol. 89, no. 1-2, pp. 5–14, 2004.

35. T. Paulmier and L. Fulcheri, "Use of non-thermal plasma for hydrocarbon reforming," Chemical Engineering Journal, vol. 106, no. 1, pp. 59–71, 2005.

36. L. Bromberg, D. R. Cohn, A. Rabinovich, C. O'Brien, and S. Hochgreb, "Plasma reforming of methane," Energy and Fuels, vol. 12, no. 1, pp. 11–18, 1998.

37. M. F. Demirbas, "Hydrogen from various biomass species via pyrolysis and steam gasification processes," Energy Sources A, vol. 28, no. 3, pp. 245–252, 2006.

38. M. Asadullah, S. I. Ito, K. Kunimori, M. Yamada, and K. Tomishige, "Energy efficient production of hydrogen and syngas from biomass: development of low-temperature catalytic process for cellulose gasification," Environmental Science and Technology, vol. 36, no. 20, pp. 4476–4481, 2002.

39. G. Weber, Q. Fu, and H. Wu, "Energy efficiency of an integrated process based on gasification for hydrogen production from biomass," Developments in Chemical Engineering and Mineral Processing, vol. 14, no. 1-2, pp. 33–49, 2006.

40. M. Ni, D. Y. C. Leung, M. K. H. Leung, and K. Sumathy, "An overview of hydrogen production from biomass," Fuel Processing Technology, vol. 87, no. 5, pp. 461–472, 2006.

41. N. Muradov, "Emission-free fuel reformers for mobile and

portable fuel cell applications," Journal of Power Sources, vol. 118, no. 1-2, pp. 320–324, 2003.

42. A. Demirba and G. Arin, "Hydrogen from biomass via pyrolysis: relationships between yield of hydrogen and temperature," Energy Sources, vol. 26, no. 11, pp. 1061–1069, 2004.

43. A. Demirba , "Recovery of chemicals and gasoline-range fuels from plastic wastes via pyrolysis,"Energy Sources, vol. 27, no. 14, pp. 1313–1319, 2005.

44. F. G. Zhagfarov, N. A. Grigor‹Eva, and A. L. Lapidus, "New catalysts of hydrocarbon pyrolysis,"Chemistry and Technology of Fuels and Oils, vol. 41, no. 2, pp. 141–145, 2005.

45. R. Sakurovs, "Interactions between coking coals and plastics during co-pyrolysis," Fuel, vol. 82, no. 15–17, pp. 1911–1916, 2003.

46. A. Ori ák, L. Halás, I. Amar, J. T. Andersson, and M. Ádámová, "Co-pyrolysis of polymethyl methacrylate with brown coal and effect on monomer production," Fuel, vol. 85, no. 1, pp. 12–18, 2006.

47. R. R. Davda, J. W. Shabaker, G. W. Huber, R. D. Cortright, and J. A. Dumesic, "Aqueous-phase reforming of ethylene glycol on silica-supported metal catalysts," Applied Catalysis B, vol. 43, no. 1, pp. 13–26, 2003.

48. G. W. Huber and J. A. Dumesic, "An overview of aqueous-phase catalytic processes for production of hydrogen and alkanes in a biorefinery," Catalysis Today, vol. 111, no. 1-2, pp. 119–132, 2006.

49. G. Eggleston and J. R. Vercellotti, "Degradation of sucrose, glucose and fructose in concentrated aqueous solutions under constant pH conditions at elevated temperature," Journal of Carbohydrate Chemistry, vol. 19, no. 9, pp. 1305–1318, 2000.

50. B. M. Kabyemela, T. Adschiri, R. M. Malaluan, and K. Arai, "Glucose and fructose decomposition in subcritical and supercritical water: detailed reaction pathway, mechanisms, and kinetics," Industrial and Engineering Chemistry Research, vol. 38, no. 8, pp. 2888–2895, 1999. ·

51. R. D. Cortright and L. Bednarova, "Hydrogen Generation from

Biomass-Derived Carbohydrates via Aqueous Phase Reforming (APR) Process," in U.S. Department of Energy Hydrogen Program FY2007 Anual Progress Report, J. Milliken, Ed., pp. 56–59, Department of Energy, Washington, DC, USA, 2007.

52. R. R. Davda, J. W. Shabaker, G. W. Huber, R. D. Cortright, and J. A. Dumesic, "A review of catalytic issues and process conditions for renewable hydrogen and alkanes by aqueous-phase reforming of oxygenated hydrocarbons over supported metal catalysts," Applied Catalysis B, vol. 56, no. 1-2, pp. 171–186, 2005.

53. B. Rozmiarek, "Hydrogen generation from biomass-derived carbohydrates via aqueous phase reforming process," in Annual Merit Review Proceedings, J. Milliken, Ed., pp. 1–6, Department of Energy, Washington, DC, USA, 2008.

54. J. Turner, G. Sverdrup, M. K. Mann et al., "Renewable hydrogen production," International Journal of Energy Research, vol. 32, no. 5, pp. 379–407, 2008.

55. S. A. Grigoriev, V. I. Porembsky, and V. N. Fateev, "Pure hydrogen production by PEM electrolysis for hydrogen energy," International Journal of Hydrogen Energy, vol. 31, no. 2, pp. 171–175, 2006.

56. J. Pettersson, B. Ramsey, and D. Harrison, "A review of the latest developments in electrodes for unitised regenerative polymer electrolyte fuel cells," Journal of Power Sources, vol. 157, no. 1, pp. 28–34, 2006.

57. C. N. Hamelinck and A. P. C. Faaij, "Future prospects for production of methanol and hydrogen from biomass," Journal of Power Sources, vol. 111, no. 1, pp. 1–22, 2002.

58. S. E. Lindquist and C. Fell, "Fuels-hydrogen production: photoelectrolysis," in Encyclopedia of Electrochemical Power Sources, G. Jürgen, Ed., pp. 369–383, Elsevier, Amsterdam, The Netherlands, 2009.

59. A. Steinfeld, "Solar thermochemical production of hydrogen—a review," Solar Energy, vol. 78, no. 5, pp. 603–615, 2005.

60. J. E. Funk, "Thermochemical hydrogen production: past and present," International Journal of Hydrogen Energy, vol. 26, no. 3, pp. 185–190, 2001.

61. M. A. Lewis, M. Serban, and J. K. Basco, "Hydrogen production at <550°C using a low temperature thermochemical cycle," in

Proceedings of the Atoms for Prosperity: Updating Eisenhower's Global Vision for Nuclear Energy (Global ‹03), pp. 1492–1498, Chicago, Ill, USA, November 2003.

62. H. Balat and E. Kirtay, "Hydrogen from biomass—present scenario and future prospects," International Journal of Hydrogen Energy, vol. 35, no. 14, pp. 7416–7426, 2010.

Chapter 8

Chemical Pretreatment Methods for the Production of Cellulosic Ethanol: Technologies and Innovations

Edem Cudjoe Bensah[1,2,3] and Moses Mensah[2]

[1]Department of Chemical Engineering, Kumasi Polytechnic, Kumasi, Ghana

[2]Department of Chemical Engineering, Kwame Nkrumah University of Science and Technology, Private Mail Bag, University Post Office, KNUST, Kumasi, Ghana

[3]Centre for Energy, Environment and Sustainable Development (CEESD), Kumasi, Ghana

ABSTRACT

Pretreatment of lignocellulose has received considerable research globally due to its influence on the technical, economic and environmental sustainability of cellulosic ethanol production. Some of the most promising pretreatment methods require the application of chemicals such as acids, alkali, salts, oxidants, and solvents. Thus, advances in research have enabled the development and integration of chemical-based pretreatment into proprietary ethanol production technologies in several pilot and demonstration plants globally, with potential to scale-up to commercial levels. This paper reviews known and emerging chemical pretreatment methods, highlighting recent findings and process innovations developed to offset inherent challenges via a range of interventions, notably, the combination of chemical pretreatment with other methods to improve carbohydrate preservation, reduce formation of degradation products, achieve high sugar yields at mild reaction conditions, reduce solvent loads and enzyme dose, reduce waste generation, and improve recovery of biomass components in pure forms. The use of chemicals such as ionic liquids, NMMO, and sulphite are promising once challenges in solvent recovery are overcome. For developing countries, alkali-based methods are relatively easy to deploy in decentralized, low-tech systems owing to advantages such as the requirement of simple reactors and the ease of operation.

INTRODUCTION

Cellulosic or second generation (2G) bioethanol is produced from lignocellulosic biomass (LB) in three main steps: pretreatment, hydrolysis, and fermentation. Pretreatment involves the use of physical processes (e.g., size reduction, steaming/boiling, ultrasonication, and popping), chemical methods (e.g., acids, bases, salts, and solvents), physicochemical processes (e.g., liquid hot water and ammonium fibre explosion or AFEX), biological methods (e.g., white-rot/brown-rot fungi and bacteria), and several combinations thereof to fractionate the lignocellulose into its components. It results in the disruption of the lignin seal to increase enzyme access to holocellulose [1, 2], reduction of cellulose crystallinity [3, 4], and increase in the surface area [5,

6] and porosity [7, 8] of pretreated substrates, resulting in increased hydrolysis rate. In hydrolysis, cellulose and hemicelluloses are broken down into monomeric sugars via addition of acids or enzymes such as cellulase. Enzymatic hydrolysis offers advantages over acids such as low energy consumption due to the mild process requirements, high sugar yields, and no unwanted wastes. Enzymatic hydrolysis of cellulose is affected by properties of the substrate such as porosity, cellulose fibre crystallinity, and degree of polymerization, as well as lignin and hemicellulose content [9, 10]; optimum mixing [11]; substrate and end-product concentration; enzyme activity; reaction conditions such as pH and temperature [12, 13]. The cost of commercial enzymes is a major economic headache in 2G bioethanol production and such pretreatment methods that support low enzyme dosages per unit biomass while optimizing ethanol yields (in addition to other favourable factors) are of interest in cellulosic ethanol production. It is known that cellulase loadings of less than 10 FPU/gram-cellulose are essential for economic production of cellulosic ethanol [14].

In fermentation, sugars are converted into ethanol under liquid- or solid-state using yeast or bacteria. The process economics is improved significantly if both C5 and C6 sugars are utilized, though the fermentation efficiency of C5 sugars is very low. Further, the yeasts cannot endure low pH as well as high ethanol and byproduct concentrations [15]. Additional challenges with xylose fermenting yeasts include long fermentation periods, low productivity, high viscosity of fermentation broth, and byproduct formation [15]. For both enzymatic hydrolysis and fermentation, the presence of degradation compounds such as furfural and hydroxymethyl-furfural (HMF) produced during pretreatment inhibits the smooth functioning of enzymes. Thus, ongoing research under pretreatment has focussed on the development of innovative methods requiring the use of mild conditions that significantly reduce inhibitor formation while maintaining high sugar yields.

In general, pretreatment presents the most practical and economic challenges in the attempt to commercialize cellulosic bioethanol [16–18] since it may affect upstream [14] as well as downstream processes by determining fermentation toxicity, enzymatic hydrolysis rates, enzyme loadings, product concentrations and purification, waste treatment demands, and power generation [19]. The results from many studies have shown that pretreatment is relatively costly among the various operations and processes involved in cellulosic ethanol production [20–25], representing about 20% of the total cost [14]. While pretreatment introduces additional cost, the consequence of hydrolysing lignocellulose without pretreatment is far less favourable since only about 20% of native biomass is hydrolysed [26].

It is generally accepted that efficient pretreatment should avoid size reduction and use of costly chemicals [19, 27], improve fibre reactivity and maximize formation/recovery of sugars [28], avoid loss of carbohydrate [29], avoid formation of enzyme-inhibiting byproducts [30, 31], preserve cellulose and hemicellulose fractions that are easily digestible by hydrolytic enzymes [32, 33], generate high-value lignin coproduct [18, 26], minimize energy requirement [22, 34], and achieve high sugar yields under high biomass loads [35]. However, no perfect pretreatment method has been discovered since there are variations in terms of suitability of one method for various materials, which may be further compounded by factors such as maturity, mode of harvest, extent of drying, and storage conditions of the feedstock [36–39]. The chemical methods of pretreating lignocellulosic materials are widely employed in many pilot and demonstration plants since they are ideal for low lignin materials. This paper reviews the various chemical methods that have received significant attention globally, with emphasis on process innovations and interesting findings from the work of researchers in the field. It is the second of two papers (with the first article addressing physical and physicochemical methods) that is expected to serve as a reference for researchers in both academia and industry. The methods discussed are centred on the use of acids, bases, oxidants, and solvents.

CHEMICAL PRETREATMENT OF LIGNOCELLULOSIC BIOMASS

Acid Hydrolysis

In this method, dried biomass is milled, (occasionally) presoaked in water, and submerged in acidic solution under specific temperatures for a period of time. Pretreated content is filtered to separate the liquor from the unhydrolysed solid substrate which undergoes washing (to extract sugars and remove acids) and/or neutralization before saccharification. Generally, the hydrogen ion concentration is directly correlated with the hydrolysis reaction constant; thus, the more negative the pKa value of the acid, the more effective the hydrolysis process [40]. Sulphuric (H_2SO_4) and phosphoric (H_3PO_4) acids are widely used since they are relatively cheap and efficient in hydrolysing lignocellulose, though the latter gives a milder effect and is more benign to the environment. Hydrochloric (HCl) acid is more volatile and easier to recover and attacks biomass better than H_2SO_4 [41]; similarly, nitric acid (HNO_3) possesses good cellulose-to-sugar conversion rates [42]. However, both acids are expensive compared to sulphuric acid.

In acidic media, the amorphous hemicelluloses in LB hydrolyse quicker (than cellulose) to soluble sugars [43, 44] and some oligomers especially in mild conditions [21] through the disruption of xylosidic bonds and cleavage of acetyl ester groups [45, 46], and the lignin seal is degraded through substitution reactions and broken links accompanied by condensation reactions that prevent dissolution [47, 48]. Cellulose undergoes preferential degradation of amorphous regions leading to enlarged cellulose fibrils and fibril aggregates [44] and an increase in the crystallinity index of the pretreated material [38, 43].

The process is generally affected by particle size, temperature, reaction time, acid concentration, and liquid-to-solid ratio. The combined severity factor (R'_0) is an index used to assess the effect of pretreatment temperature, time, and pH on the efficiency of the process as shown below [49]:

$$\log R'_0 = \log \left(t \cdot e^{(T-100)/14.75}\right) - \text{pH}, \tag{1}$$

where t is the reaction time in minutes and T is the hydrolysis temperature in °C. Acid pretreatment comes under two main variations—dilute and concentrated acid hydrolysis, each applied under further process variations.

Dilute Acid Pretreatment

Under dilute acid (0.2–2.5% w/w) processes, high temperatures (120–210°C) and pressures are used to achieve reaction times in seconds or minutes and are thus suitable for continuous operations [21, 50]. The low acid consumption is a major advantage in terms of cost and process severity [51]. Moreover, low acid concentrations (<1% w/v sulphuric/phosphoric) release essential nutrients (S and P) that enhance downstream fermentation [52]. A variation of the process involves two stages of pretreatment: in the first stage, most hemicelluloses in the biomass substrate are solubilised in the presence of a more dilute acid, while the second stage involves the use of a higher acid concentration to hydrolyse the cellulose and the remaining hemicellulose [53, 54].

Increasing pretreatment severity generally leads to higher rates of cellulose conversion to glucose at shorter reaction times, though conditions that yield high glucose do not necessarily translate to high xylose yields. Pretreatment of municipal solid waste (carrot and potato peelings, grass, newspaper, and crap paper) with dilute acid H_2SO_4, HNO_3, and HCl was undertaken by Li et al. [55] who found the glucose yield of pretreated substrates to depend more on acid concentration and enzyme loading than reaction temperature. For feedstock mixtures such as aspen and switchgrass, and aspen and balsam, the process was observed to have no synergistic or antagonistic effect on enzymatic hydrolysis, indicating the likelihood of predicting such combined systems based on models for pure species yields [56]. A comparatively low number of researchers have investigated the nitric acid pretreatment of LB. Dilute HNO_3 pretreatment was found to give the highest glucose concentration (compared to dilute H_2SO_4) in the pretreatment of rye straw [57]. However, byproducts from nitric acid pretreatment are difficult to remove by washing of the pretreated substrates [42].

Also, dilute H_3PO_4 is frequently used; the acid was applied on potato peels with overall sugar yield reaching 82.5% of the theoretical even though arabinose conversion was found to be low due to its thermal instability [58]. Its application to bamboo and corn cob also yielded high sugars at 170°C for 45 minutes [59] and 140°C for 10 minutes, respectively [60]. In another work, Avci et al. [61] achieved 85% glucose and 91.4% xylose yields on corn stover at 0.5% (v/v)/180°C/15 min and 1% (v/v)/160°C/10, respectively, at low concentrations of degradation products. In an attempt to improve process performance, researchers have tried combinations of acids and other compounds with mixed results. Heredia-Olea et al. [62] combined HCl and H_2SO_4 in pretreating sweet sorghum bagasse but did not record any significant improvement over acid treatment. However, Zhang et al. [52] recorded higher xylose yields during combined (H_2SO_4/H_3PO_4) pretreatment of oil palm empty fruit bunch compared to single acid application. Anaerobic storage of pretreated substrates was found to improve enzymatic degradability of reed canarygrass and switchgrass in a pilot-scale plant [63].

Aside mineral acids, organic acids such as maleic, fumaric [8], and oxalic [64–67] have been found to be useful substitutes to mineral acids. Oxalic and maleic acids have higher solution potential and degrade hemicelluloses more efficiently than sulphuric acid [8]. In addition, they have two pKa values which favour efficient hydrolysis over a range of temperatures and pH values [66, 68]. The use of maleic and fumaric generally results in lower degradations products at similar conditions compared to sulphuric acid [69]. Oxalic acid application to corn stover gave best results at 160°C for 10 minutes at a concentration of 200 mM [70]. On maple wood, oxalic pretreatment resulted in equivalent glucose (87.7%), xylose (86.9%), and total sugar yields (87.4%) compared to dilute sulphuric and hydrochloric acids [71].

Main Disadvantages. Though dilute acid pretreatment has received wide attention from researchers due to its advantages such as high cellulose content of pretreated substrates and low requirement of enzymes, it comes with some drawbacks. A major demerit of this process is its requirement of special corrosion-resistant reactors which are usually expensive both in investment and operation [72], compared to other chemical (e.g., dilute alkali) and physicochemical (e.g., steam explosion and AFEX) methods [13, 73]. The energy consumption of the process and the cost of the acid [74] as well as performance limitations based on particle size (a few millimetres) and solids concentration ($\leq$30%) contribute significantly to the overall cost [14]. Dilute acids are less effective in removing lignin compared to alkaline methods. Neutralization of pretreated contents creates solid waste [75], though it is necessary for improving the downstream fermentation process.

The sugar yield is reduced because a portion of the sugar is degraded into enzyme-inhibiting byproducts such as furfural (2-furaldehyde), 5-hydroxymethylfurfural (5-HMF), acetic acid, gypsum, vanillin, and aldehydes (4-hydroxybenzaldehyde, syringaldehyde, etc.) as a result of the conditions that cause cellulose to rapidly break into sugars [76–79]. It is pertinent that inhibitors are removed by filtering off hydroxylate liquor followed by washing and drying of cellulose-rich residues [80] or by using reverse osmosis to exclude acetic acid, furfural 5-HMF, and other compounds before fermentation [81, 82]. The use of membranes to detoxify hydroxylates could be exploited at the industrial level due to the ease of scaling up for large-scale operations. Further, employing simultaneous saccharification and fermentation (SSF) ensures rapid conversion of glucose into ethanol and the continuous removal of ethanol during fermentation [83]. Other options pertain to the use of agents such as activated carbon to selectively absorb inhibitors [84] or the use of yeast strains that tolerate inhibitors at significant high levels [85].

Another compound known to cause inhibition of enzymatic hydrolysis of cellulose is pseudolignin—an acid-insoluble substance that is formed from repolymerization of degradation products with/without lignin. Surface lignin reduces cellulase affinity for biomass substrate and lowers the hydrolysis rate [86]. High acid concentrations, temperatures, and treatment time generally create conditions for increased degradation products and pseudolignin, resulting in reduced lignin recovery in the hydroxylate. Inhibitor concentrations may be reduced by combining mild physical-chemical conditions and optimised enzyme loadings [87], by using a two-stage process [53] or by performing deacetylation prior to pretreatment to reduce hydroxylate toxicity [45]. Losses in C5 sugars solubilised into the liquid stream are avoided if it is further processed [33] or if unfiltered hydroxylates are wholly used in enzymatic hydrolysis, where neutralization of the hydroxylate can be achieved by the use of a novel single-step neutralization and buffering procedure to adjust the pH [87, 88].

Concentrated Acid Pretreatment

This pretreatment variation uses concentrated sulphuric (65–86% w/v), hydrochloric (41%), or phosphoric (85% w/w) acids to pretreat dried (5–10% moisture), pulverized biomass at low temperatures (30–60°C) and pressures. Pretreated contents are diluted with deionized water for saccharification to take place at moderate temperatures (70–121°C), separated into solid and liquid fractions, followed by washing and neutralization of the solid substrates [89, 90]. Another variation involves the addition of organic solvents such as acetone to pretreated contents followed by agitation of the mixture to stop the reaction and to separate the solids from the supernatant (containing lignin) which undergo further washing before enzymatic hydrolysis [91].

The efficiency of the process is affected by acid concentration, acid/biomass ratio, process temperature, and time. Two patented process configurations—the Arkenol and the Biosulfurol processes—which are based on the concentrated H_2SO_4 platform have shown potential commercially. The Arkenol process involves pretreatment (decrystallization of biomass) at temperatures below 50°C at acid (70–77%)/biomass (10% moisture, <1 mm particle size) ratio of 1.25 [92]. In the Biosulfurol process, the biomass is trickled in the acid (70%) in the presence of dry CO_2 from the fermenter followed by dilution of the pretreated slurry with water. The acid is partly separated from the biomass slurry by the use of membranes before fermentation and partly in an anaerobic digester after fermentation. The relative merits of the Biosulfurol process, according to van Groenestijn et al. [92], include the nonrequirement of enzymes, low temperature treatment, low production of degradation products, and the capacity to fractionate various biomass with high ethanol yields.

Concentrated H_3PO_4 is effective at low temperatures, dissolves cellulose in the presence of water, possesses no inhibitory effects on hydrolysis and fermentation, and gives high sugars [93] compared to dilute acid pretreatment [89]. Pretreated biomass substrates are left with uneven and rough molecular surfaces that enhance enzyme adsorption rates and thereby accelerate hydrolysis [94]. Combinations of acid and p-cresol, a phenol derivative, enabled complete separation of lignin and carbohydrates in the pretreatment of oil palm empty fruit bunch [90]. In cellulose solvent- and organic solvent-based lignocelluloses fractionation (COSLIF), the biomass feedstock is fractionated by adding concentrated H_3PO_4 under mild conditions (50°C, 1 atm, and 60 min) followed by the addition of an organic solvent such as ethanol (95% v/v) or acetone under room temperature for 10 min [95, 96]. The main advantages of COSLIF include high enzyme accessibility to cellulose in pretreated substrates [96] and effective fractionation of diverse biomass with good sugar yields [95]. High glucan digestibility at low enzyme dose is usually achieved with COSLIF [96,97] and in combination with other agents such as ionic liquids [97].

Despite the aforementioned merits, COSLIF (like other concentrated acid methods) is slow and uses high loads of solvent. Further, depolymerisation and sugar degradation increase at temperatures above 50°C [95]. Other challenges with concentrated acid pretreatment include corrosion of equipment, acid recovery, and neutralization waste when acid is not recovered.

Alkali Pretreatment

Process Background

In alkaline pretreatment, lignocellulosic materials are mixed with bases such as sodium, potassium, calcium, and ammonia [50, 98] at specific temperatures and pressures in order to degrade ester and glycosidic side chains of the material [18], leading to lignin structure disruption [99, 100], cellulose swelling, and decrystallization [74]. Alkaline treatment extracts hemicelluloses from polysaccharides and produces organic acids that lower the pH. Two streams are formed comprising a wet solid fraction composed of mainly cellulose and a liquid fraction containing dissolved hemicelluloses, lignin, and some unreacted inorganic chemicals. The solids are separated and washed in warm/hot water until neutrality before they are hydrolysed. Washing removes enzyme inhibitors and residual unreacted reagents and improves the release of sugars from pretreated solids.

The process is influenced by NaOH loading, liquid-to-solid ratio, temperature, and time, among others. In general, pretreatment is less severe since it can be carried out under atmospheric conditions though at the cost of longer retention times [50, 101]. Low alkali concentrations (<4% w/w) are mostly used at high temperatures and pressures. Mild alkali pretreatment of biomass favours enzymatic hydrolysis especially for materials that have relatively low lignin content. Sugar degradation and corrosion problems are less severe in alkali processes than in acid pretreatment [57, 102] and mild conditions (55°C) may not require posttreatment washing since enzyme inhibiting compounds are generally low [103].

Bases that have been used widely include hydroxides of potassium, sodium, and calcium, as well as sodium carbonate (Na_2CO_3). KOH selectively removes xylan [104] as was observed on peashrub at 25°C for 10 h with high efficiency [105]. Application of KOH pretreatment on rye straw gave lower sugar yields than dilute acid [57]. It was, however, found to give high fermentation efficiency relative to other methods [42]. Regarding Na_2CO_3, efficient delignification of biomass is realized which makes the carbonate a promising chemical for pretreatment. Application of the carbonate [106] as well as combined Na_2CO_3-Na_2SO_3 [107] on rice straw recorded good carbohydrate preservation and sugar yields.

The bases—NaOH and $Ca(OH)_2$—have been extensively investigated as agents for pretreatment and are discussed in the sections below.

Sodium Hydroxide Pretreatment

Though expensive, sodium hydroxide (NaOH) is used widely due to its relative high alkalinity for the fractionation of various materials including agricultural residues and wood. Dilute NaOH application loosens the biomass structure, separate the bonds between the lignin and the carbohydrates, increases the internal surface area, decreases the degree of polymerization and crystallinity, and disrupts the lignin structure [108]. Homogenization of pretreated substrates enhances glucose yields by increasing the surface area and porosity of the biomass [6]. High alkaline concentrations generally cause increased biomass delignification [109], though severe concentrations (6–20% w/w) result in cellulose dissolution and reduced lignin removal [110]. Pedersen et al. [33] observed that increasing the pH from 10 to 13 increased the removal of lignin from 40 to 80% w/w dry wheat straw at 140°C. Also, by varying the NaOH loading rate from 3 to 9% based on initial dry bagasse, Zhao et al. [111] recorded an increase in delignification, from 52.3% to 75.5%.

Compared to acid hydrolysis, NaOH pretreatment appears to improve enzymatic biodegradability due to the higher delignification ability of alkali. Evidence of this assertion, for example, can be found in the work of Ioelovich and Morag [112] who applied both dilute H_2SO_4 and NaOH to four materials and found the alkali to be more efficient in terms of sugar yields, delignification, and biomass utilization rate.

In many situations, positive results have come out from combined NaOH and other agents such us peracetic acid [111], polyelectrolyte [113], and hydrogen peroxide. In addition, the environmental impact is low and no special reactors are required. As a pretreatment agent, alkaline peroxide favours enzymatic hydrolysis as it removes lignin efficiently [114] and produces insignificant inhibitors [115, 116]. Its application to corn stover was effective in producing good glucose yields at low (0.125 g H_2O_2/g biomass) peroxide doses [117]. Further, higher sugar yields from grass stovers were obtained using peroxide relative to dilute NaOH and AFEX [118]. Another positive aspect about peroxide is its use in recovering lignin and other components from substrates pretreated by other methods which results in higher sugar yields [116].

Lime Pretreatment

Lime—in the form of quick lime, CaO, and slaked lime, $Ca(OH)_2$—has also been extensively investigated as a pretreatment agent due to its low cost, safety in handling, availability in many countries, and ease of recovery. In pretreatment, lime and water are added to the feedstock at temperatures ranging from ambient to 130°C, sometimes in the presence of oxygen to enhance delignification [21, 119]. A loading of 0.1 gram of slake lime per gram of biogas is common and process time varies from hours to weeks [26]. Lime improves hydrolysis rates of biomass by removing acetyl groups and a considerable portion of the lignin fraction [120], reducing counter-productive cellulase adsorption [14] and formation of degradation byproducts [121], and promoting cellulose accessibility [122].

Lime has been applied on various feedstocks with encouraging results. Lignin was selectively removed at low carbohydrate losses in the treatment of sugarcane bagasse at optimised conditions of 90°C for 90 h at lime loading of 0.4 g/g bagasse [123]. Lime application to feedstocks such as, inter alia, corn stover [124], switchgrass [125], and sugarcane bagasse [126] gave high carbohydrate conversions to simple sugars. Combined lime and oxidants so far have shown positive results [119] and further investigations are needed. The ease with which lime can be recovered—via precipitation to $CaCO_3$ using CO_2—is a major advantage. The carbonate may be combusted alongside with lignin and other residues in a boiler to ash and CaO which can be mixed with water and slaked to form $Ca(OH)_2$ [21].

Despite the merits of lime pretreatment, its use faces drawbacks such as longer reaction period (influenced by the reaction temperature) compared to NaOH under similar conditions. Moreover, it dissolves in water at a slower rate and thus requires higher volumes of water for pretreatment. In some instances, lime pretreatment produced substrates with less favourable characteristics; for example, lime pretreated sugarcane grass was less amenable to cellulose hydrolysis compared to dilute acid [78].

General Drawbacks of Alkaline Methods

Alkaline pretreatment is generally unsuitable for woody biomass due to the requirement of severe conditions needed to fractionate recalcitrant wood. Thus, lignin removal may be improved via methods that include oxygenation [19] and the addition of chemicals such as urea [108]. Other drawbacks are the loss of hemicelluloses and the formation of inhibitors at harsh conditions. Further, the formation of salts upon neutralization of pretreated contents may present challenges with disposal. Salts hamper the purification of pretreated hydroxylates [127], while posttreatment washing results in sugar losses [103]. Higher catalyst loadings are generally used and thus require recovery and reuse to improve process economics in an industrial scale plant [128].

Wet Oxidation (WO)

Process Description

In WO, biomass undergoes oxidation in an aqueous (acidic, neutral, and alkaline conditions) solution via reaction with oxygen (air) at elevated temperatures (125–315°C) and pressures (0.5–5 MPa) [54, 129]. The pretreated suspension is filtered to separate the cellulose-rich solid from the hemicellulose-rich filtrate, and the solid component is washed with deionized water before undergoing enzymatic hydrolysis. WO pretreatment oxidizes the hemicellulose fractions of materials into intermediates such as carboxylic acids—via peeling reactions and chain cleavage, and from the phenolic structures of lignin [130], acetaldehydes, and alcohols, and finally to CO_2 and H_2O [131]. The degree of fractionation, in most cases, is influenced by the reaction temperature more than the time and oxygen dose [138]. High temperatures, pressures, pH, and catalysts favour rapid oxidation [129]; the catalysts further cause increases in the acid (formic and oxalic) concentration regardless of the reaction temperature [130].

Generally, alkaline WO reduces the formation of enzyme-inhibiting compounds such as furfural and HMF compared to acidic and neutral conditions [132, 133]. High phenolic concentrations reduce the volumetric productivity of the enzyme [134] by causing partition and loss of integrity of cell membranes of the fermenting organisms and are thus more toxic than HMF and furfural [7]. Reports from several researchers indicate that wet oxidation achieves good hydrolysis and fermentation yields from various LB, notably, spruce [130, 131], wheat straw [134], rape straw [135], and rice husk [136]. Under optimized conditions of 185°C, 5 bar, and 15 min, about 67% of cellulose in the solid fraction of rice husk was obtained, while 89 and 70% of lignin and hemicelluloses were removed [136]. WO pretreatment (195°C, 15 min, 12 bar, 2 g/L Na_2CO_3) and subsequent enzymatic hydrolysis of winter rye, oilseed rape, and faba bean produced ethanol yields of 66%, 70%, and 52% of the theoretical, respectively [137].

Drawbacks

WO is costly to operate owing to the need to supply high pressure oxygen and chemicals such as sodium carbonate [138]. It appears that alkaline WO does not favour woods as was observed in the pretreatment of spruce [131] and willow [133], where optimized conditions were found at 12 bar and 200°C in 10 min; and 185°C, 12 bar O_2, 15 min, respectively, under neutral conditions. One potential option for reducing cost is to use air instead of oxygen in a modified process known as wet air oxidation (WAO) as was employed for the pretreatment of shea-tree sawdust [139], resulting in maximum sugar yield of 263.5 mg glucose/g dry biomass at optimum conditions of 150°C/45 min/1% H_2O_2/10 bar air [140].

Organosolv Pretreatment

Process Description

The organosolv process involves the addition of an (aqueous) organic solvent mixture with/without a catalyst—such as an acid (HCl, H_2SO_4, etc.), a base (e.g., NaOH), or a salt ($MgCl_2$, $Fe_2(SO_4)_3$, etc.)—to the biomass under specific temperatures and pressures [13, 141, 142]. The process produces three main fractions—a high purity lignin, a hemicellulosic syrup containing C5 and C6 sugars, and a relatively pure cellulose fraction. The pretreated solid residues are separated by filtration and washed with distilled water to remove solvents and degradation products which may possess inhibitory characteristics to downstream process such as enzymatic hydrolysis and fermentation. Pretreatment conditions lead to simultaneous hydrolysis and lignin removal [74] via the disruption of internal bonds in lignin, lignin-hemicellulose bonds, and glycosidic bonds in hemicelluloses and to a smaller extent in cellulose [142]. Other changes include the formation of droplets of lignin on the surface of pretreated biomass, a situation that inhibits hydrolysis by adsorption on the surface of the cellulose [143]. Process variables such as temperature, reaction time, solvent concentration, and acid dose affect the physical characteristics (crystallinity, degree of polymerization of cellulose, and fibre length) of the pretreated substrates. In most situations, high temperatures and acid concentrations as well as elongated reaction times cause considerable degradation of sugar into fermentation inhibitors.

Though sulphuric acid has been used extensively as a catalyst due to its strong reactivity, it is toxic and corrosive and possesses inhibitory characteristics [144]. Park et al. [144] evaluated the effectiveness of acidic (H_2SO_4), basic (NaOH), and neutral ($MgCl_2$) catalysts on pine and found the acid as the most efficient in terms of the ethanol yield; however, an increase in the concentration of the base from 1 to 2% had a positive effect on digestibility. Organic solvents/acids that have been used as catalysts include formic, oxalic, acetylsalicylic, salicylic acid, methanol, ethanol, acetone, ethylene glycol, triethylene glycol, and tetrahydrofurfuryl alcohol [13, 145–147]. The use of CO_2 as a catalyst did not improve process yields on pretreatment of willow wood [151].

Process Variations

Organosolv pretreatment comes in several variations based on the solvent type, catalyst, and process conditions. The Battelle organosolv method involves the use of a ternary mixture of phenol, water, and HCl to fractionate the biomass at about 100°C and at 1 atm [148]. The acid depolymerizes lignin and hydrolyses the hemicellulose fraction and the lignin dissolves in the organic phase (phenol), while the monosaccharides are attracted to the aqueous phase upon cooling of the fractionated biomass.

Similarly, the formic acid organosolv process (formasolv) involves the application of formic acid, water, and HCl to depolymerize, oxidize, and dissolve lignin, hemicellulose, and extractives in the biomass, and the precipitation of the lignin is achieved by the addition of water [148]. Formic acid has a good lignin solvency and the process can be undertaken under low temperatures and at atmospheric pressure [147]. However, formic acid may cause formylation of the pretreated substrates which could reduce cellulose digestibility. Pretreated substrates can be deformylated in alkaline solution as was observed on bagasse at 120°C [147]. Formic acid (5–10% w/w; no catalysts) was used by Kupiainen et al. [149] on delignified wheat straw pulp at 180–220°C, yielding a maximum glucose yield of 40%.

Organosolv pretreatment with acetic acid (acetosolv) produces higher yields than formic since less mass is dissolved for a given time. In addition, acetosolv achieves higher cellulose viscosity in smaller time periods [148].

The use of ethanol in organosolv pretreatment (ethanosolv) enables the recovery of high value products including cellulose, sulphur- and chlorine-free lignin, enriched hemicelluloses, and extractives. Further purification may be achieved via the use of solvents such as ionic liquids [127]. Unlike formasolv, the ethanosolv process is usually operated under higher pressures and temperatures. In addition, reprecipitation of lignin occurs due to lower lignin solubility [147]. The reduced toxicity of ethanol—compared to solvents such as methanol—to the downstream fermentation process and the fact that ethanol is the final product are additional benefits [142, 150]. Generally, lower ethanol/water ratios favour hemicellulose hydrolysis and enzymatic degradability of pretreated substrates [151] since ethanol is an inhibitor to the performance of hydrolytic enzymes. Ethanosolv has been explored for the development of proprietary technologies such as the Alcell process which is a sustainable alternative to kraft pulping [152], and the Lignol process—a biorefinery platform that uses aqueous ethanol (50% w/w) for pretreating LB at 200°C and 400 psi to separate the various components in woody biomass [153, 154]. High sugar yields and product recovery have been observed on ethanosolv pretreatment of various materials including hybrid poplar [155] and Japanese cypress [156]. A major advantage is the potential to recover much of the ethanol [143, 157] and water [141] which reduces the operating cost. Ethanosolv coproducts such as hemicellulose syrup and lignin can serve as feedstocks for the production of high value biochemicals. Moreover, ethanosolv lignins whose functional groups and molecular weight depend on process conditions are known to possess antioxidant properties [158].

In several situations, presoaking materials, for example, in acidic medium [159] or in bioslurry [160], positively affect the process in terms of sugar yields and lignin removal, among others. Other variations involve the combined use of acid and basic catalysts [146], microwave-assisted organosolv [161], and the avoidance of catalysts [162]. In another variation, the biomass is treated with the inclusion of ferric sulphate and sodium hydroxide to the biomass/liquor (formic acid and hydrogen peroxide) mixture. Formic acid reacts with hydrogen peroxide to produce peroxyformic acid and its application in organosolv pretreatment (Milox) of biomass produced good results on both hardwoods and softwoods [148].

Drawbacks

Though organosolv is promising due to the potential to obtain byproducts in pure forms for the manufacture of high-value biochemicals, the process is generally costly to operate due to the requirement of high temperatures and pressures. The use of mineral acids in the organosolv process is an environmental concern, and corrosion due to the use of organic acids is a challenge. In addition, pretreated substrates need washing to prevent lignin from precipitating, and recovery of expensive volatile organic solvents needs very efficient control systems and additional energy requirements [163].

Ionic Liquids (Green Solvents)

Properties of Ionic Liquids

Ionic liquids (ILs) are salts consisting of large cations (mostly organic) and small anions (mostly inorganic), with a low degree of cationic symmetry and a melting point below 100°C. ILs are nonflammable [14], are liquid at room temperature [164], and are known to improve antielectrostatic and fire-proof properties of wood [165]. They have low volatility and high thermal stability [14] up to temperatures of about 300°C [166], high electrical conductivity, high solvating properties, and wide electrical window [167]. Other favourable aspects involve characteristics such as water stability, polarity, refractive index, and density [168].

Ionic liquids exist in two main forms—simple salts comprising single cations and anions, and those where equilibrium is involved [50]. The most common forms contain the imidazolium cation which can pair with anions such as chloride, bromide, acetate, sulphate, nitrate, methanoate, and triflate. ILs could be designed and developed to pretreat specific biomass under optimal conditions by combining cations and anions which can result in an estimated formulation of 10^9 ILs [169]. For example, the ILs 1-ethyl-3-methylimidazolium glycinate (Emim-Gly) and 1-allyl-3-methylimidazolium chloride (Amim-Cl) were synthesized from various compounds for the dissolution of bamboo [170] and wood [71], respectively. In addition, the properties of ILs could be altered by varying the length and branching of the alkyl groups that are integrated into the cation [171].

Not all characteristics of ionic liquids are favourable as solvents in pretreatment. For example, chloride-based ILs such as 1-butyl-3-methylimidazolium chloride (Bmim-Cl) are toxic, corrosive, and very hygroscopic, while others such as Amim-Cl are viscous with reactive side chains [172]. Also, ILs with long akyl chains have the tendency to obstruct nonpolar active sites of enzymes due to their hydrophobic nature [173]. Others have favourable properties and have thus been under intense investigation as promising solvents. For example, phosphate-based solvents possess higher thermal stability and lower viscosity and toxicity than chloride-based ones [167].

Positive outcomes have also been recorded with the use of 1-ethyl-3-methylimidazolium acetate (Emim-Ac) since it is favourable to in situ enzymatic saccharification due to its biocompatibility and enzymatic activity [174].

Process Description and Mechanism

In IL pretreatment, a mixture of the biomass (0.1–0.5 mm) and the solvent—sometimes in the presence of water and acid—is incubated at temperatures ranging from 80 to 160°C for 10 minutes to 24 hours, followed by the addition of an antisolvent to precipitate the cellulose fraction. In acidic conditions, biomass dissolution is followed by acid hydrolysis of dissolved cellulose [175]. The pretreated supernatant is removed via centrifugation or filtration, and the cellulose is washed with distilled water, lyophilized (freeze-dried), and saccharified. The antisolvent is separated from the IL by processes such as flash distillation as the IL is recovered for reuse. ILs convert carbohydrates in lignocellulosic materials into fermentable sugars via two main pathways: one is the pretreatment of the biomass to improve its enzymatic hydrolysis efficiency, and the other focuses on the transformation of the hydrolysis process from a heterogeneous to a homogeneous reaction system by dissolution in the solvent [176].

Both the cation and the anion function differently in the dissolution of biomass [169]. The effectiveness of pretreatment is predicted using the Kamlet-Taft hydrogen bond acceptor ability, β: usually, the higher values of β translate to higher lignin removal and vice versa [177]. It is known that component ions in ILs influence enzyme activity [178] and stability [173], and the anion—as a result of its hydrogen basicity—attacks and breaks the hydrogen bonds of cellulose structure [179] by forming hydrogen bonds with the cellulose [180]. Thus, the cellulose is solubilised and the crystalline nature is reduced [181] via cell wall swelling resulting from dislocation of hydrogen bonds between cellulose fibrils and lignin [182], partial removal of hemicellulose, and biomass delignification [183].

The dissolved cellulose in IL-pretreated hydroxylate is precipitated (regenerated) into cellulose II on addition of anti-solvents such as water [182, 184], methanol [184, 185], acetone [186], and ethanol [166] via preferential solute-displacement mechanism. The type of antisolvent was found to have no effect on the structure of regenerated cellulose in the pretreatment of Avicel, according to Dadi et al. [185]. Regenerated cellulose is more amorphous and has additional sites for enzyme adsorption which is favourable to saccharification. Addition of water in the cause of IL-biomass reaction improves cellulose hydrolysis by increasing selectivity to glucose and cellobiose [187]. If antisolvent mixtures such as water/acetone are used, then cellulose and lignin are distinctly separated through dissolution in water and acetone, respectively [170,188].

The temperature is a key parameter that influences sugar release pattern, saccharification kinetics, and sugar yields [189]. Higher temperatures and pretreatment times are more efficient in solubilising lignin [190]. From the work of Zhi-Guo and Hong-Zhang [191], nearly 100% increase in glucose yield was recorded on pretreatment of wheat straw with Amim-Cl when the temperature was increased from 125 to 150°C at a reaction time of 2 h. At ambient conditions, a fraction cellulose I may remain and recrystallize to microfibrils of cellulose I upon expulsion of the solvent [192].

Application

ILs are increasingly being used to dissolve various LB as shown in Table 1. The effectiveness of the solvent depends partly on the type of biomass being pretreated and the final application of pretreated substrates including the regenerated cellulose. Carbohydrate losses are generally low and degradation products are significant only at severe conditions. Through the work of authors such as Cheng et al. [193], Xie et al. [172], Dadi et al. [166], Hou et al. [194], and others, it is known that significant sugar yields are achieved without the elimination of crystallinity.

Table 1: Application of ILs to selected biomass resources

Biomass	Solvent	Temperature (°C), time (min)	Reference
Eucalyptus	Emim-Ac	150, 3	[183]
	Amim-Cl	120, 5	[184]
Poplar	Emim-Ac	120, 1	[195]
Pine	Amim-Cl	120, 5	[184]
Spruce	Amim-Cl	120, 5	[184]

Energy cane bagasse	Emim-Ac	120, 0.5	[196]
Switchgrass	Emim-Ac	160, 3	[38]
		120–160, 6	[190]
Bamboo	Emim-Gly	120, 8	[170]
Wheat straw	Amim-Cl	100–150, 2–6	[191]
	Bmim-Ac	100–150, 0.17–1	[164]
Water hyacinth	Bmim-Ac	100–150, 0.17–1	[164]
Rice husk	Bmim-Cl, Emim-Ac	100, 10	[197]
Rice straw	Ch-Aa	90, 2	[194]
Kenaf powder	Ch-Ac	110, 16 h	[198]
Cassava pulp	Emim-Ac, Dmim-SO$_4$, Emim-DePO$_4$	25–120, 24	[188]

Dmim-SO$_4$: 1, 3-Dimethylimidazolium methyl sulphate. Emim-DePO$_4$: 1-Ethyl-3-methylimidazolium diethyl phosphate. Ch-Aa: Cholinium amino acids; Ch-Ac: Cholinium acetate.

Drawbacks and Process Modifications

Though the use of ionic liquids is under consideration for large scale applications, several challenges such as high solvent cost, high solvent loading, technical challenges and cost of solvent regeneration, and inadequate knowledge on the impact of ionic liquids on the environmental [38, 165, 167]. ILs with high viscosities have low potential in terms of mass and phase transfer which presents challenges in engineering applications [165, 175]. Moreover, the separation of hydrophilic ILs and monomeric sugars in water is difficult [178, 204]. Some ILs also exhibit tendencies to denature enzymes [168] and the active sites of enzymes could be blocked by layers of hydrophilic ILs, decreasing or destroying the aqueous phase surrounding enzyme surface [173]. Washing of regenerated cellulose as well as recycling of ILs via processes such as evaporation and reverse osmosis is practically costly which presents challenges in the development of process technologies for the efficient use of ILs within a biorefinery. While delignification increases with high temperatures, such conditions also cause hemicellulose losses as was observed in the pretreatment of switchgrass and agave bagasse at 120 and 160°C [190].

In order to address some of the shortcomings of conventional IL pretreatment approach, several process routes and configurations are being developed. The use of lower IL concentrations is realized by using aqueous ILs [175] or by undertaking both pretreatment and saccharification in a single unit followed by direct extraction of sugars which avoids the need to wash regenerated cellulose. Aqueous ILs have lower viscosities, reduce recycling demands in terms of cost, and effectively deal with high biomass loadings [175]. Emerging green solvents such as cholinium-based ILs have been found to be more biocompatible and renewable and lower in cost, and with yields comparable to imidazolium-based ILs [198].

Economic improvements have been reported via interventions such as the use of thermophilic cellulase in high solvent concentrations [205]; application of acid catalysts such as sulfonated carbon materials (sugar catalysts) [206], HCl [175, 187], and Nafion NR50 [207]; inclusion of chemicals such as NH_4OH-H_2O_2 [208] and electrolyte solution [181]; direct (without cellulase) conversion of pretreated substrates to biofuels in consolidated bioprocessing [209]; the reuse of solvent in several batches [169]. Notwithstanding other process enhancements, it is necessary that lignin-derived products are efficiently utilized to improve process economics.

Oxidizing Agents

Unlike the methods discussed above, pretreatment involving the application of oxidizing agents has received less attention among researchers partly due to the high cost of the oxidants such as ozone and hydrogen peroxide. The sections below discuss oxidant-based methods, highlighting recent advances in research and process development.

Ozonolysis

In ozonolysis, ozone is sparged into a mixture of biomass and water at room temperature and specific time periods leading to the solubilisation of lignin and hemicelluloses. The focus of attack is on the aromatic ring of lignin [7], and the process is affected by ozone concentration, biomass type and moisture content, and air/ozone flow rate [210]. Though the process is relatively expensive due to large requirements of ozone, the process comes with benefits as follows: high dry matter concentrations (45–60%), effective removal of lignin, very low production of inhibitory products, and reactions performed at atmospheric conditions [211, 212]. Ozone, thus, barely attacks carbohydrates [212]. Lignin degradation products such as carboxylic acids that may form can be eliminated by washing with water at room temperature even though it comes at the expense of some carbohydrates losses.

In some instances, ozone application to specific biomass resulted in low sugar yields. In basic medium, ozone application was found to be inefficient on wheat and rye straw [210]. Ozone treatment of cotton stalk (10% w/v) at 4°C for 30–90 minutes reduced lignin by 11.97–16.6%, at xylan and glucan solubilisation of 1.9–16.7% and 7.2–16.6%, respectively; comparatively, NaOH treatment achieved higher delignification of 65.63% [109]. In another work, high delignification and low carbohydrate loss were observed when a two-step method comprising ozone and ethanosolv was applied to Sweetgum, Miscanthus, and Loblolly pine [213]. In addition, combined ozonolysis and autohydrolysis offer benefits such as high hemicellulose solubilisation, high glucose and ethanol yields, low use of chemicals, and low waste production [213].

Recently, the use of plasma-generated ozone (from air or oxygen-enriched air) at atmospheric conditions has attracted interest among researchers including Schultz-Jensen and team at the Risø National Laboratory for Sustainable Energy in Denmark [211]. Employing a fixed-bed reactor, a CO_2 detector, and a technique for continuous determination of ozone consumption, lignin degradation of ozone pretreatment of wheat straw was monitored in real time with respect to ozone consumption and CO_2 emission. Lignin degradation of 1 mm particles was found to be almost complete while that of 2 mm particles was less than 80%, leaving a solid fraction mainly composed of carbohydrates. Maximum glucose and ethanol yields of 78% and 52% were observed after enzymatic hydrolysis and SSF (based on glucan), respectively, based on optimal ozonisation for 1 h [214]. The ethanol yield was relatively low and the process economics has the potential to be improved via the recovery and use of lignin byproducts and hemicellulose, as well as developing schemes to reduce ozone consumption for similar yields.

Aside high delignification and carbohydrate preservation, pretreated substrates have shown potential for use in the production of enzymes by fungi such as Trichoderma reesei, with lower titres of cellulases and higher amount of xylanases recorded from autoclave sterilization of pretreated materials, compared to nonsterilized substrates with antibiotics added [215].

Other Oxidizing Agents

When exposed to biomass, oxidants such as hydrogen peroxide (H_2O_2), peracetic acid ($C_2H_4O_3$), sulphur trioxide (SO_3), and chlorine dioxide (ClO_2) solubilise hemicellulose and lignin under mild alkaline or neutral conditions. The pretreated biomass undergoes oxidative delignification due to the reaction of the aromatic ring of lignin with the oxidizing agents leading to improved digestibility compared to alkaline pretreatment only [7]. Biomass degradability is affected by the type of biomass, oxidant dose, reaction temperature, and time. Among the oxidants, hydrogen peroxide is the most studied. Hydrogen peroxide degrades into hydrogen and oxygen and does not leave residues in the biomass [216]. By pretreating water hyacinth and lettuce with NaOH followed by H_2O_2, Mishima et al. [217] recorded higher sugar yields compared to NaOH, H_2SO_4, and hot water pretreatments under similar conditions. Process variations that have also produced high sugar yields include the addition of catalysts such as manganese acetate [218], postpretreatment acid saccharification [216], and alkaline-peroxide application without postpretreatment washing [219].

Generally, H_2O_2 permits fractionation of biomass at ambient pressures and low temperatures, allowing the use of low cost reactors. Unfortunately, the application of oxidizing agents produces soluble lignin compounds that inhibit the conversion of hemicelluloses and cellulose to ethanol. There is also loss of sugar due to the occurrence of nonselective oxidation [7, 104]. In addition, the high cost of oxidants is a major limitation for scaling up to industrial levels.

Recently, the use of sulphur trioxide in a process called sulphur trioxide microthermal explosion (STEX) has been explored to pretreat biomass such as rice straw [220]. Biomass is hanged above a solution of oleum (50% SO_3) and NaOH (1% w/v) and swirled in a test tube at 50°C/1 atm for 7 h, followed by washing to obtain the solids [220]. The internal explosion is believed to occur due to heat released from SO_3-straw reaction that causes rapid expansion of air, and water from the interior of the biomass, resulting in enhanced structural changes and pore volume [220] and partial removal of lignin and hemicellulose [221]. Pretreatment of rice straw at the aforementioned conditions resulted in saccharification yield of 91% [220]. The efficient handling of SO_3 will be a challenge due its corrosiveness regarding this emerging pretreatment. Another novel process involves the use of chlorine dioxide (in the presence of aqueous ethanol) on biomass which results in high fraction of glucose in pretreated biomass and low formation of inhibiting compounds [222].

Sulphite Pretreatment

The sulphite process is a matured technology that has been used in the pulp and paper industry for decades and has been adapted to the pretreatment of LB for enhanced enzymatic hydrolysis and fermentation to ethanol. In this method, milled biomass (<6 mm) is mixed in sulphite (Na_2SO_3, $NaHSO_3$, etc.) solution (1–10% w/w) in acidic, basic, or neutral environments at selected temperatures (80–200°C) and reaction times (30–180 min). Pretreated solids are separated from the spent liquor by filtration and the solids are washed with distilled water and dried before undergoing saccharification. The process partially degrades and sulphonates lignin and enhances glucose yields due to the formation of sulphonic and weak acid groups which improves the hydrophilicity of pretreated substrates [223]. Sulphonation is enhanced in the presence of volatile organic solvents such as ethanol which reduces the surface tension thereby allowing effective solution penetration of the biomass. In addition, the lignin is hydrolysed and dissolved in the organic phase and is easily recovered in pure forms [223]. Lignin removal can be accelerated further by the presence of sulphomethyl groups produced from combined action of sulphite and formaldehyde on lignin, resulting in high sugar yields [224].

Sulphite application is emerging as a promising pretreatment due to positive results recorded from several materials. Pretreatment of corn stover with alkaline Na_2SO_3 at 140°C resulted in 92% lignin removal and 78.2% total sugar yield (0.48 g/g raw biomass) after enzymatic hydrolysis, which was higher than four other alkali-based methods under similar conditions [225]. About 79.3% of total glucan was converted to glucose and cellobiose during corn cob pretreatment at 156°C, 1.4 h, 7.1% charge, and solids loading of 1 : 7.6 w/w; and subsequent SSF gave 72.2% theoretical ethanol yield [226]. In another work, pretreatment of corn cob residues produced the highest glucose yields (81.2%) in the presence of ethanol compared to acidic, basic, and neutral conditions [223].

At the pilot level, sulphite-based methods have been investigated and demonstrated good results. In a proprietary process, known as sulphite pretreatment to overcome recalcitrance of lignocellulose (SPORL), the biomass material is first pretreated with dilute solutions of sulphuric acid and sodium bisulphite ($NaHSO_3$) after which the residual solids are separated from the hydroxylate. The solids are then disc-milled and pressed to obtain solids moisture content of about 30% [227]. Lignin is sulphonated and partially removed, and hemicellulose is nearly removed completely which favours subsequent size reduction and enzymatic hydrolysis [228]. The SPORL process has been effectively used to pretreat (180°C, 25 min, liquor/wood = 3 : 1 v/w) lodgepole pine for subsequent conversion to ethanol at a yield of 276 L/ton wood and at a net energy output of 4.55 GJ/ton wood [227]. SPORL pretreatment of switchgrass was superior to dilute acid [228, 229] and alkali [228] in terms of the digestibility of the pretreated substrates. Similarly, higher sugar yields and lower inhibitor concentration were found with SPORL-pretreated agave stalk relative to dilute acid and NaOH [230]. It was also found superior to the organosolv and steam explosion pretreatments based on the evaluation of total sugar recovery and energy consumption [23]. The main advantages of sulphite pretreatment are high sugar yields, effective lignin removal, and recovery of biomass components in less chemically transformed forms. The drawbacks include sugar degradation at severe conditions, large volumes of process water used in postpretreatment washing, and the high costs of recovering pretreatment chemicals. It has, however, been shown through the work of Liu and Zhu [231] that the negative effects of soluble inhibitors and lignosulphonate on enzymatic hydrolyses could be counteracted by adding metal salts to the pretreated contents, making it possible to avoid the costly washing process.

Glycerol

Crude glycerol—a byproduct in biodiesel production—has attracted interest as a substrate for fermentation into ethanol and other biochemicals [232] and also as a solvent for fractionating biomass in order to improve the economics of cellulosic ethanol as well as the upstream biodiesel production. Glycerol pretreatment causes efficient delignification of biomass. Guragain et al. [164] investigated the best conditions in the use of crude glycerol (water : glycerol = 1 : 1) in pretreating wheat straw and water hyacinth and arrived at 230°C for 4 h for wheat straw and 230°C for 1 h for water hyacinth. Enzymatic hydrolysis of pretreated wheat straw produced reducing sugar yields (mg/g of sample) of 423 and 487 for crude and pure glycerol, respectively, compared to a low figure of 223 for dilute acid. In addition, hydrolysis tests on water hyacinth gave yields of 705, 719, and 714 for crude glycerol, pure glycerol, and dilute acid, respectively. Similarly, Ungurean et al. [186] performed glycerol pretreatment of wood (poplar, acacia, oak, and fir) and recorded higher cellulose conversion rates compared to dilute acid application. However, combinations of glycerol and acid/IL pretreatment yielded higher sugar levels compared to glycerol pretreatment alone.

There are wide variations in the composition of crude glycerol which usually contains methanol, ash, soap, catalysts, salts, and nonglycerol organic matter, among others, in varied proportions [233]. While the potential for exploring crude glycerol application together with other methods is high, there is the need to assess the quality of crude glycerol and its effects on sugar and ethanol yields of promising feedstocks.

Aqueous N-Methylmorpholine-N-Oxide (NMMO)

NMMO is a well-known industrial solvent used in the Lyocell process for the production of fibres and has attracted interest for use as a pretreatment solvent. Cellulose dissolves (without derivatization) in NMMO/H_2O system and the hydrogen bonds in cellulose are disrupted in favour of new bonds between cellulose and solvent molecules [234], leading to swelling and increased porosity [235], as well as reduced degree of polymerization and crystallinity which improves enzymatic saccharification. Addition of boiled distilled water to pretreatment slurry containing dissolved biomass causes cellulose I to precipitate into cellulose II which is more reactive. Regenerated solids are filtered and washed with warm/boiling water until the filtrate is clear.

Like ionic liquids, NMMO dissolves biomass with no/less chemical modification at low/moderate temperatures (80–130°C). Additional favourable characteristics of NMMO pretreatment include high sugar yields, formation of low degradation products, high solvent recovery, and no adverse effect on the environment. Further, cellulase activity is not negatively affected by low concentrations (15–20% w/w) of NMMO, indicating the potential of application in continuous processes [235].

Aqueous NMMO has already being used on a host of biomass as a sole pretreatment method or in combination with others. Almost total conversion of cellulose to ethanol at ethanol yields of up to 85.4% and 89% for pretreatment (130°C/3 h) of oak and spruce, respectively, was observed [236]. High saccharification yields (>90%) were observed for ultrasound-assisted NMMO treatment of sugarcane bagasse [233], as well as pretreatment of birch [237]. Poornejad et al. [238] compared the effectiveness of NMMO and the ionic liquid (Bmim-Ac) treatment on rice straw at 120°C/5 h. Glucan conversion was complete with the ionic acid, while 96% conversion was realized with the use of NMMO. Upon SSF, the yield of ethanol was higher with NMMO (93.3%) than Bmim-Ac (79.7%).

At the pilot scale, concentrated NMMO pretreatment (85% w/w, 130°C, 5 h) has been investigated on spruce and birch by Lennartsson et al. [239]. For wood chips below 2 mm, maximum hydrolysis yields (mg/g wood) ranged from 195 to 128 for spruce and 136 to 175 for birch depending on the scale of the pilot reactor using nonisothermal SSF. A technoeconomic analysis of NMMO pretreatment of spruce for ethanol and biogas production was undertaken by Shafiei et al. [240] who observed relatively high process energy efficiency of 79%.

Inorganic Salts

Recently, salts such as iron (III) chloride [241] and calcium chloride [242] have found use in the pretreatment of biomass. The salt hydrolyses in water to form a strong acidic solution which causes rapid removal of hemicelluloses from biomass during pretreatment.

Application of the salt (0.26 M) to olive tree residues at 152.6°C for 30 min resulted in 100% removal of hemicellulose, and subsequent saccharification produced a yield of 36.6 g glucose/100 g of glucose in the original biomass [241]. Combined $CaCl_2$-microwave treatment was investigated on corn stover and under optimum conditions (162.1°C, 12 min); glucose recovery reached 65.5% which was higher than that of steam-exploded (1.5 MPa, 5 min) corn stover [242].

FUTURE OUTLOOK AND CONCLUSION

Due to its high reactivity at mild conditions, chemical pretreatment of lignocellulosic biomass forms the basics of several proprietary cellulosic ethanol production configuration and technologies that have been developed by various research groups and companies for development at various levels, usually with financial support from national governments and public bodies (e.g., Swedish Energy Agency, Danish Ministry of Energy, US Department of Energy/Agriculture, and Canadian Sustainable Development Technology Canada) and multinational institutions such as the European Union. Table 2 gives profiles of some of the main projects undertaken or under construction/development underpinned by breakthrough pretreatment, hydrolysis, and fermentation technologies, as well as process integration and optimization.

Table 2: Selected large-scale cellulosic ethanol plants based on chemical pretreatment

Company/institution	Location	Capacity	Pretreatment	Feedstock	Status	Reference
SEKAB	Örnsköldsvik, Sweden	160 t/y	Two-stage dilute H_2SO_4/SO_2	Pine chips	Operational since 2004; scale-up projects planned	[138] Updated
Abengoa bioenergy	Salamanca, Spain	5 mL/y		Wheat and barley straw, and so forth	Operational since 2009	http://www.km.aiche.org/mange-bwter_p9/prometations-E1_Babb.pdf (accessed 11/02/13)
	York, NE USA	0.4 mL/v	Acid impregnation + steam explosion	Corn stover	Operational since 2008	
	Kansas, USA	75 m gal/y		Corn stover, wheat straw, switchgrass, and so forth	Under construction. Startup 2014+	
BioGasol	Ballerup, Denmark	4 t/h	Dilute acid, steam explosion or wet explosion	Wheat straw and bran, corn stover, and so forth, garden wastes, energy crops, and green wastes	Under development	http://www.biogasol.com/Plants-3.aspx (accessed 11/02/13)

Name	Location	Capacity	Process	Feedstock	Status	Reference
Procethol 2G, Futurol	Pomacle, France	180 m³/y 3.5 mL/y 180 mL/y	Under development	Wheat straw, switchgrass, green waste, Miscanthus, Vinasses, and so forth	Operational since 2011 Planned 2015 Planned 2016	http://www.projet-futurol.com/index-uk.php (accessed 13/02/13)
Izumi Biorefinery	Japan	300 L/day	Arkenol	Cedar, pine, and hemlock	Operational since 2002	[200]; http://bfreinc.com/docs/IZUMI_Status_2004_for_BlueFire_051606.pdf (accessed 13/02/13)
INEOS	Florida, USA	8 m gal/y	Thermochemical	Municipal solid waste and so forth	Operational	http://www.ineosbio.com/80-Technology_platform.htm (accessed 11/02/13)
ZeaChem	Oregon, USA	250,000 gal/y	Chemical	Hybrid poplar, corn stover, and cob	Construction completed	http://www.zeachem.com/index.php (accessed 12/02/13)
Logos Technologies	California, USA	50,000 gal/y	Colloid milling	Corn stover, switchgrass, and wood chips	Operational	http://www.logos-technologies.com/ (accessed 13/02/13)
BlueFire	Mississippi, USA	19 m gal/y	Concentrated acid (Arkenol)	Wood waste, municipal solid waste	Under construction	http://bfreinc.com/ (accessed 15/02/13)
Weyland AS	Norway	200 m³/y	Concentrated acid	Corn stover, sawdust, paper pulp, switchgrass, and so forth	Operational since 2010	http://weyland.no/ (accessed 15/02/13)
Borregaard	Norway	20 mL/y	Acidic/neutral sulphite	Wheat straw, eucalyptus, spruce, and so forth	Operational	[201]; http://www.borregaard.com/ (accessed 11/02/13)
Queensland University of Technology	Queensland, Australia	Pilot	Acid, alkaline, steam explosion, ionic liquid, and so forth	Sugarcane bagasse	Operational since 2010	[202]
Praj Industries	India	10 mL/y	Thermochemical	Corn cob, sugarcane bagasse, and so forth	Under development	[206] (Updated)
Lignol Energy Corporation	Canada	100 m³/y 40–200 m L/y	Organosolv	Wood, agricultural residues, and so forth	Operational Planning	http://www.lignol.ca/index.html (accessed 06/02/13)
Blue sugars	Wyoming, USA	1.4 m gal/y	Acid, thermomechanical	Pine and so forth	Operational	http://bluesugars.com/ (accessed 13/02/13)
Petrobras/Blue sugars	Brazil	10 m gal/y	Acid, thermomechanical	Sugarcane bagasse	Under development	http://bluesugars.com/company-partners.htm (accessed 18/02/13)
Dupont Danisco Cellulosic Ethanol (DDCE)	Iowa, USA	30 m gal/y	NH₃ Steam recycled	Corn stover	Under construction	[300]; http://biofuels.dupont.com/ (accessed 13/02/13)
COFCO /SINOPEC /Novozyme	Zhaodong, China	62 mL/y	Steam explosion (with acid impregnation)	Corn stover	Under construction	[203]

*Pilot (<2000 t/y or 2.5 m m³/y), demo (2000–50,000 t/y or 2.5–65 m m³/y), commercial (>50,000 t/y or 65 m m³/y).

The pretreatment of feedstocks to enhance biodegradability to simple sugars has been the subject of intensive research globally with a focus on maximizing sugar yields at high solid loads and at the lowest economic and environmental costs. This paper has reviewed widely known and emerging chemical pretreatment methods with regard to process description, advantages, drawbacks, and recent innovations employed to offset inherent challenges. Though cellulosic ethanol is close to commercialization, there are still technical, economic, and environmental challenges associated with biomass pretreatment, hydrolysis, and fermentation. No solvent has been found to work best for all biomass and such optimized methods and process conditions for various materials need to be investigated and developed further.

Some major challenges of the chemical pretreatment include the requirement of extensive size reduction; handling biomass at high solids concentration, corrosion, solvent costs, and recovery; environmental pollution from solvents, byproducts, and waste from reactions. Nonetheless, the aforementioned challenges are being dealt with via several interventions, notably, the application of novel solvents and the combination of different chemical methods with physico-chemical and biological pretreatments to achieve higher sugar yields, milder process conditions, lower use of costly solvents, low enzyme loads, recovery, and use of biomass components in pristine forms, and improvements in environmental sustainability.

ACKNOWLEDGMENTS

The authors are grateful to DANIDA through the development research project (DFC journal no. 10-018RISØ) titled—Biofuels production from lignocellulosic materials (2GBIONRG).

REFERENCES

1. J. S. Lim, Z. Abdul Manan, S. R. Wan Alwi, and H. Hashim, "A review on utilisation of biomass from rice industry as a source of renewable energy," Renewable and Sustainable Energy Reviews, vol. 16, no. 5, pp. 3084–3094, 2012.

2. Y. Z. Pang, Y. P. Liu, X. J. Li, K. S. Wang, and H. R. Yuan, "Improving biodegradability and biogas production of corn stover through sodium hydroxide solid state pretreatment," Energy and Fuels, vol. 22, no. 4, pp. 2761–2766, 2008.
3. J. Gabhane, S. P. M. Prince William, A. N. Vaidya, K. Mahapatra, and T. Chakrabarti, "Influence of heating source on the efficacy of lignocellulosic pretreatment—a cellulosic ethanol perspective,"Biomass and Bioenergy, vol. 35, no. 1, pp. 96–102, 2011.
4. Y. Kim, R. Hendrickson, N. S. Mosier et al., "Enzyme hydrolysis and ethanol fermentation of liquid hot water and AFEX pretreated distillers› grains at high-solids loadings," Bioresource Technology, vol. 99, no. 12, pp. 5206–5215, 2008.
5. J.-S. Lee, B. Parameswaran, J.-P. Lee, and S.-C. Park, "Recent developments of key technologies on cellulosic ethanol production," Journal of Scientific and Industrial Research, vol. 67, no. 11, pp. 865–873, 2008.
6. Y. Li, R. Ruan, P. L. Chen et al., "Enzymatic hydrolysis of corn stover pretreated by combined dilute alkaline treatment and homogenization," Transactions of the American Society of Agricultural Engineers, vol. 47, no. 3, pp. 821–825, 2004.
7. P. Harmsen, W. Huijgen, L. Bermudez, and R. Bakker, "Literature review of physical and chemical pretreatment processes for lignocellulosic biomass," Tech. Rep. 1184, Biosynergy, Wageningen UR Food & Biobased Research, 2010.
8. J.-W. Lee and T. W. Jeffries, "Efficiencies of acid catalysts in the hydrolysis of lignocellulosic biomass over a range of combined severity factors," Bioresource Technology, vol. 102, no. 10, pp. 5884–5890, 2011.
9. J. S. Van Dyk and B. I. Pletschke, "A review of lignocellulose bioconversion using enzymatic hydrolysis and synergistic cooperation between enzymes-Factors affecting enzymes, conversion and synergy," Biotechnology Advances, vol. 30, no. 6, pp. 1458–1480, 2012.
10. J. D. McMillan, "Bioethanol production: status and prospects," Renewable Energy, vol. 10, no. 2-3, pp. 295–302, 1997.
11. D. M. Lavenson, E. J. Tozzi, N. Karuna, T. Jeoh, R. L. Powell, and M. J. McCarthy, "The effect of mixing on the liquefaction and

saccharification of cellulosic fibers," Bioresource Technology, vol. 111, pp. 240–247, 2012.

12. G. Radeva, I. Valchev, S. Petrin, E. Valcheva, and P. Tsekova, "Kinetic model of enzymatic hydrolysis of steam-exploded wheat straw," Carbohydrate Polymers, vol. 87, no. 2, pp. 1280–1285, 2012.

13. Y. Sun and J. Cheng, "Hydrolysis of lignocellulosic materials for ethanol production: a review,"Bioresource Technology, vol. 83, no. 1, pp. 1–11, 2002.

14. B. Yang and C. E. Wyman, "Pretreatment: the key to unlocking low-cost cellulosic ethanol," Biofuels, Bioproducts and Biorefining, vol. 2, no. 1, pp. 26–40, 2008.

15. R. C. Kuhad, R. Gupta, Y. P. Khasa, A. Singh, and Y.-H. P. Zhang, "Bioethanol production from pentose sugars: current status and future prospects," Renewable and Sustainable Energy Reviews, vol. 15, no. 9, pp. 4950–4962, 2011.

16. R. Choudhary, A. L. Umagiliyage, Y. Liang, T. Siddaramu, J. Haddock, and G. Markevicius, "Microwave pretreatment for enzymatic saccharification of sweet sorghum bagasse," Biomass and Bioenergy, vol. 39, pp. 218–226, 2012.

17. M. Galbe and G. Zacchi, "Pretreatment: the key to efficient utilization of lignocellulosic materials,"Biomass Bioenergy, vol. 46, pp. 70–78, 2012.

18. N. Sarkar, S. K. Ghosh, S. Bannerjee, and K. Aikat, "Bioethanol production from agricultural wastes: an overview," Renewable Energy, vol. 37, no. 1, pp. 19–27, 2012.

19. C. E. Wyman, B. E. Dale, R. T. Elander, M. Holtzapple, M. R. Ladisch, and Y. Y. Lee, "Coordinated development of leading biomass pretreatment technologies," Bioresource Technology, vol. 96, no. 18, pp. 1959–1966, 2005.

20. T. Eggeman and R. T. Elander, "Process and economic analysis of pretreatment technologies,"Bioresource Technology, vol. 96, no. 18, pp. 2019–2025, 2005.

21. L. Tao, A. Aden, R. T. Elander et al., "Process and technoeconomic analysis of leading pretreatment technologies for lignocellulosic ethanol production using switchgrass," Bioresource Technology, vol. 102, no. 24, pp. 11105–11114, 2011.

22. C. E. Wyman, B. E. Dale, R. T. Elander et al., "Comparative sugar recovery and fermentation data following pretreatment of poplar wood by leading technologies," Biotechnology Progress, vol. 25, no. 2, pp. 333–339, 2009.
23. J. Y. Zhu and X. J. Pan, "Woody biomass pretreatment for cellulosic ethanol production: technology and energy consumption evaluation," Bioresource Technology, vol. 101, no. 13, pp. 4992–5002, 2010.
24. S. Sánchez-Segado, L. J. Lozano, A. P. de los Ríos, F. J. Hernández-Fernández, C. Godinez, and D. Juan, "Process design and economic analysis of a hypothetical bioethanol production plant using carob pod as feedstock," Bioresource Technology, vol. 104, pp. 324–328, 2012.
25. J. Littlewood, R. J. Murphy, and L. Wang, "Importance of policy support and feedstock prices on economic feasibility of bioethanol production from wheatstraw in the UK," Renewable & Sustainable Energy Reviews, vol. 17, pp. 291–300, 2013.
26. N. Mosier, C. Wyman, B. Dale et al., "Features of promising technologies for pretreatment of lignocellulosic biomass," Bioresource Technology, vol. 96, no. 6, pp. 673–686, 2005.
27. D. Chiaramonti, A. M. Rizzo, M. Prussi, et al., "2nd generation lignocellulosic bioethanol: is torrefaction a possible approach to biomass pretreatment?" Biomass Conversion and Biorefinery, vol. 1, pp. 9–15, 2011.
28. T. Rogalinski, T. Ingram, and G. Brunner, "Hydrolysis of lignocellulosic biomass in water under elevated temperatures and pressures," Journal of Supercritical Fluids, vol. 47, no. 1, pp. 54–63, 2008.
29. G. Hu, J. A. Heitmann, and O. J. Rojas, "Feedstock pretreatment strategies for producing ethanol from wood, bark, and forest residues," BioResources, vol. 3, no. 1, pp. 270–294, 2008.
30. V. B. Agbor, N. Cicek, R. Sparling, A. Berlin, and D. B. Levin, "Biomass pretreatment: fundamentals toward application," Biotechnology Advances, vol. 29, no. 6, pp. 675–685, 2011.
31. I. F. Cullis and S. D. Mansfield, "Optimized delignification of wood-derived lignocellulosics for improved enzymatic hydrolysis," Biotechnology and Bioengineering, vol. 106, no. 6, pp. 884–893, 2010.

32. S. Thongkheaw and B. Pitiyont, Enzymatic Hydrolysis of Acid-Pretreated Sugarcane Shoot, vol. 60, World Academy of Science, Engineering and Technology, 2011.
33. M. Pedersen, K. S. Johansen, and A. S. Meyer, "Low temperature lignocellulose pretreatment: effects and interactions of pretreatment pH are critical for maximizing enzymatic monosaccharide yields from wheat straw," Biotechnology for Biofuels, vol. 4, article 11, 2011.
34. J. Y. Zhu, X. Pan, and R. S. Zalesny Jr., "Pretreatment of woody biomass for biofuel production: energy efficiency, technologies, and recalcitrance," Applied Microbiology and Biotechnology, vol. 87, no. 3, pp. 847–857, 2010.
35. C. M. Roche, C. J. Dibble, and J. J. Stickel, "Laboratory-scale method for enzymatic saccharification of lignocellulosic biomass at high-solids loadings," Biotechnology for Biofuels, vol. 2, no. 1, article 28, pp. 1–11, 2009.
36. B. Bals, C. Rogers, M. Jin, V. Balan, and B. Dale, "Evaluation of ammonia fibre expansion (AFEX) pretreatment for enzymatic hydrolysis of switchgrass harvested in different seasons and locations,"Biotechnology for Biofuels, vol. 3, article 1, 2010.
37. B. S. Dien, H.-J. G. Jung, K. P. Vogel et al., "Chemical composition and response to dilute-acid pretreatment and enzymatic saccharification of alfalfa, reed canarygrass, and switchgrass," Biomass and Bioenergy, vol. 30, no. 10, pp. 880–891, 2006.
38. C. Li, B. Knierim, C. Manisseri et al., "Comparison of dilute acid and ionic liquid pretreatment of switchgrass: biomass recalcitrance, delignification and enzymatic saccharification," Bioresource Technology, vol. 101, no. 13, pp. 4900–4906, 2010.
39. Z.-H. Liu, L. Qin, M.-J. Jin, et al., "Evaluation of storage methods for the conversion of corn stover biomass to sugars based on steam explosion pretreatment," Bioresource Technology, vol. 132, pp. 5–15, 2013.
40. O. Bobleter, "Hydrothermal degradation of polymers derived from plants," Progress in Polymer Science, vol. 19, no. 5, pp. 797–841, 1994.
41. A. Demirbas, "Products from lignocellulosic materials via degradation processes," Energy Sources A, vol. 30, no. 1, pp. 27–37, 2008.

42. M. Tutt, T. Kikas, and J. Olt, "Influence of different pretreatment methods on bioethanol production from wheat straw," Agronomy Research Biosystem Engineering, vol. 1, pp. 269–276, 2012.
43. R. Samuel, Y. Pu, M. Foston, and A. J. Ragauskas, "Solid-state NMR characterization of switchgrass cellulose after dilute acid pretreatment," Biofuels, vol. 1, no. 1, pp. 85–90, 2010.
44. M. Foston and A. J. Ragauskas, "Changes in lignocellulosic supramolecular and ultrastructure during dilute acid pretreatment of Populus and switchgrass," Biomass and Bioenergy, vol. 34, no. 12, pp. 1885–1895, 2010.
45. X. Chen, J. Shekiro, M. A. Franden et al., "The impacts of deacetylation prior to dilute acid pretreatment on the bioethanol process," Biotechnology for Biofuels, vol. 5, article 8, 2012.
46. J. Tang, K. Chen, J. Xu, J. Li, and C. Zhao, "Effects of dilute acid hydrolysis on composition and structure of cellulose in eulaliopsis binata," BioResources, vol. 6, no. 2, pp. 1069–1078, 2011.
47. R. G. Candido, G. G. Godoy, and A. R. Gonçalves, Study of Sugarcane Bagasse Pretreatment With Sulfuric Acid As a Step of Cellulose Obtaining, vol. 61, World Academy of Science, Engineering and Technology, 2012.
48. S. V. Pingali, V. S. Urban, W. T. Heller et al., "Breakdown of cell wall nanostructure in dilute acid pretreated biomass," Biomacromolecules, vol. 11, no. 9, pp. 2329–2335, 2010.
49. H. L. Chum, D. K. Johnson, S. K. Black, and R. P. Overend, "Pretreatment-catalyst effects and the combined severity parameter," Applied Biochemistry and Biotechnology, vol. 24-25, pp. 1–14, 1990. ·
50. V. Menon and M. Rao, "Trends in bioconversion of lignocellulose: biofuels, platform chemicals & biorefinery concept," Progress in Energy and Combustion Science, vol. 38, no. 4, pp. 522–550, 2012.
51. M. Galbe and G. Zacchi, "A review of the production of ethanol from softwood," Applied Microbiology and Biotechnology, vol. 59, no. 6, pp. 618–628, 2002.
52. D. Zhang, Y. L. Ong, Z. Li, and J. C. Wu, "Optimization of dilute acid-catalyzed hydrolysis of oil palm empty fruit bunch for high yield production of xylose," Chemical Engineering Journal, vol. 181-182, pp. 636–642, 2012.

53. F. K. Kazi, J. Fortman, R. Anex et al., "Techno-Economic analysis of biochemical scenarios for production of cellulosic ethanol," Tech. Rep. NREL/TP-6A2-46588, NREL, 2010.
54. F. Talebnia, D. Karakashev, and I. Angelidaki, "Production of bioethanol from wheat straw: an overview on pretreatment, hydrolysis and fermentation," Bioresource Technology, vol. 101, no. 13, pp. 4744–4753, 2010.
55. A. Li, B. Antizar-Ladislao, and M. Khraisheh, "Bioconversion of municipal solid waste to glucose for bio-ethanol production," Bioprocess and Biosystems Engineering, vol. 30, no. 3, pp. 189–196, 2007.
56. M. Brodeur-Campbell, J. Klinger, and D. Shonnard, "Feedstock mixture effects on sugar monomer recovery following dilute acid pretreatment and enzymatic hydrolysis," Bioresource Technology, vol. 116, pp. 320–326, 2012.
57. M. Tutt, T. Kikas, and J. Olt, Comparison of Different Pretreatment Methods on Degradation of Rye Straw, Engineering for Rural Development, Jelgava, Latvia, 2012.
58. P. Lenihan, A. Orozco, E. O'Neill, M. N. M. Ahmad, D. W. Rooney, and G. M. Walker, "Dilute acid hydrolysis of lignocellulosic biomass," Chemical Engineering Journal, vol. 156, no. 2, pp. 395–403, 2010.
59. B. Hong, G. Xue, L. Weng, and X. Guo, "Pretreatment of moso bamboo with dilute phosphoric acid,"BioResources, vol. 7, no. 4, pp. 4902–4913, 2012.
60. S. Satimanont, A. Luengnaruemitchai, and S. Wongkasemjit, "Effect of temperature and time on dilute acid pretreatment of corn cobs," International Journal of Chemical and Biological Engineering, vol. 6, 2012.
61. A. Avci, B. C. Saha, B. S. Dien, et al., "Response surface optimization of corn stover pretreatment using dilute phosphoric acid for enzymatic hydrolysis and ethanol production," Bioresource Technology, vol. 130, pp. 603–612, 2013.
62. E. Heredia-Olea, E. Pérez-Carrillo, and S. O. Serna-Saldívar, "Effects of different acid hydrolyses on the conversion of sweet sorghum bagasse into C5 and C6 sugars and yeast inhibitors using response surface methodology," Bioresource Technology, vol. 119, pp. 216–223, 2012.

63. M. F. Digman, K. J. Shinners, R. E. Muck, and B. S. Dien, "Pilot-scale on-farm pretreatment of perennial grasses with dilute acid and alkali for fuel ethanol production," Transactions of the ASABE, vol. 53, no. 3, pp. 1007–1014, 2010.
64. J.-W. Lee, C. J. Houtman, H.-Y. Kim, I.-G. Choi, and T. W. Jeffries, "Scale-up study of oxalic acid pretreatment of agricultural lignocellulosic biomass for the production of bioethanol," Bioresource Technology, vol. 102, no. 16, pp. 7451–7456, 2011.
65. J.-W. Lee, R. C. L. B. Rodrigues, H. J. Kim, I.-G. Choi, and T. W. Jeffries, "The roles of xylan and lignin in oxalic acid pretreated corncob during separate enzymatic hydrolysis and ethanol fermentation," Bioresource Technology, vol. 101, no. 12, pp. 4379–4385, 2010.
66. X. Li, Z. Cai, E. Horn, and J. E. Winandy, "Effect of oxalic acid pretreatment of wood chips on manufacturing medium-density fiberboard," Holzforschung, vol. 65, no. 5, pp. 737–741, 2011.
67. X. Li, Z. Cai, E. Horn, and J. E. Winandy, "Oxalic acid pretreatment of rice straw particles and loblolly pine chips: release of hemicellulosic carbohydrates," Tappi Journal, vol. 10, no. 5, pp. 41–45, 2011. ·
68. D. Scordia, S. L. Cosentino, J.-W. Lee, and T. W. Jeffries, "Dilute oxalic acid pretreatment for biorefining giant reed (Arundo donax L.)," Biomass and Bioenergy, vol. 35, no. 7, pp. 3018–3024, 2011.
69. A. A. Modenbach and S. E. Nokes, "The use of high-solids loadings in biomass pretreatment-a review," Biotechnology and Bioengineering, vol. 109, no. 6, pp. 1430–1442, 2012.
70. G. Y. S. Mtui, "Oxalic acid pretreatment, fungal enzymatic saccharification and fermentation of maize residues to ethanol," African Journal of Biotechnology, vol. 11, no. 4, pp. 843–851, 2012.
71. T. Zhang, R. Kumar, and C. E. Wyman, "Sugar yields from dilute oxalic acid pretreatment of maple wood compared to those with other dilute acids and hot water," Carbohydrate Polymers, vol. 92, pp. 334–344, 2013.
72. M. Sharma and A. Kumar, "Xylanases: an overview," British Biotechnology Journal, vol. 3, no. 1, pp. 1–28, 2013.

73. D. Kumar and G. S. Murthy, "Impact of pretreatment and downstream processing technologies on economics and energy in cellulosic ethanol production," Biotechnology for Biofuels, vol. 4, article 27, 2011.
74. C. Conde-Mejía, A. Jiménez-Gutiérreza, and M. El-Halwagi, "A comparison of pretreatment methods for bioethanol production from lignocellulosic materials," Safety and Environmental Protection, vol. 90, no. 3, pp. 189–202, 2012.
75. C. Cara, I. Romero, J. M. Oliva, F. Sáez, and E. Castro, "Liquid hot water pretreatment of olive tree pruning residues," Applied Biochemistry and Biotechnology, vol. 136–140, no. 1–12, pp. 379–394, 2007.
76. Z. Anwar, M. Gulfraz, M. J. Asad et al., "Bioethanol productions from rice polish by optimization of dilute acid pretreatment and enzymatic hydrolysis," African Journal of Biotechnology, vol. 11, no. 4, pp. 992–998, 2012.
77. T. S. Jeong, C. H. Choi, J. Y. Lee, and K. K. Oh, "Behaviors of glucose decomposition during acid-catalyzed hydrothermal hydrolysis of pretreated Gelidium amansii," Bioresource Technology, vol. 116, pp. 435–440, 2012.
78. R. Jutakanoke, N. Leepipatpiboon, V. Tolieng, V. Kitpreechavanich, T. Srinorakutara, and A. Akaracharanya, "Sugarcane leaves: pretreatment and ethanol fermentation by Saccharomyces cerevisiae," Biomass and Bioenergy, vol. 39, pp. 283–289, 2012.
79. Y.-B. Yi, J.-W. Lee, Y.-H. Choi, S.-M. Park, and C.-H. Chung, "Simple process for production of hydroxymethylfurfural from raw biomasses of girasol and potato tubers," Biomass and Bioenergy, vol. 39, pp. 484–488, 2012.
80. Z. Anwar, M. Gulfraz, M. Imran, et al., "Optimization of dilute acid pretreatment using response surface methodology for bioethanol production from cellulosic biomass of rice polish," Pakistan Journal of Botany, vol. 44, no. 1, pp. 169–176, 2012.
81. D. L. Grzenia, D. J. Schell, and S. R. Wickramasinghe, "Membrane extraction for detoxification of biomass hydrolysates," Bioresource Technology, vol. 111, pp. 248–254, 2012.
82. A. M. Shupe and S. Liu, "Ethanol fermentation from hydrolysed hot-water wood extracts by pentose fermenting yeasts," Biomass and Bioenergy, vol. 39, pp. 31–38, 2012.

83. L. Viikari, J. Vehmaanpera, and A. Koivula, "Lignocellulosic ethanol: from science to industry,"Biomass Bioenergy, vol. 46, pp. 13–24, 2012.
84. Y. H. Jung, I. J. Kim, H. K. Kim, and K. H. Kim, "Dilute acid pretreatment of lignocellulose for whole slurry ethanol fermentation," Bioresource Technology, vol. 132, pp. 109–114, 2013.
85. R. Koppram, E. Albers, and L. Olsson, "Evolutionary engineering strategies to enhance tolerance of xylose utilizing recombinant yeast to inhibitors derived from spruce biomass," Biotechnology for Biofuels, vol. 5, article 32, 2012.
86. X. Ju, M. Engelhard, and X. Zhang, "An advanced understanding of the specific effects of xylan and surface lignin contents on enzymatic hydrolysis of lignocellulosic biomass," Bioresource Technology, vol. 132, pp. 137–145, 2013.
87. L. Favaro, M. Basaglia, and S. Casella, "Processing wheat bran into ethanol using mild treatments and highly fermentative yeasts," Biomass Bioenergy, vol. 46, pp. 605–617, 2012.
88. X. Gao, R. Kumar, J. D. DeMartini, H. Li, and C. E. Wyman, "Application of high throughput pretreatment and co-hydrolysis system to thermochemical pretreatment—part 1: dilute acid,"Biotechnology and Bioengineering, vol. 110, no. 3, pp. 754–762, 2013.
89. Z. S. Liu, X. L. Wu, K. Kida, and Y. Q. Tang, "Corn stover saccharification with concentrated sulfuric acid: effects of saccharification conditions on sugar recovery and by-product generation," Bioresource Technology, vol. 119, pp. 224–233, 2012.
90. M. H. M. Zainudin, N. A. Rahman, S. Abd-Aziz, et al., "Utilization of glucose recovered by phase separation system from acid-hydrolysed oil palm empty fruit bunch for bioethanol production,"Pertanika Journal of Tropical Agricultural Science, vol. 35, no. 1, pp. 117–126, 2012.
91. Y.-H. P. Zhang, S.-Y. Ding, J. R. Mielenz et al., "Fractionating recalcitrant lignocellulose at modest reaction conditions," Biotechnology and Bioengineering, vol. 97, no. 2, pp. 214–223, 2007.
92. J. W. van Groenestijn, J. H. O. Hazewinkel, and R. R. Bakker,

Pre-Treatment of Ligno-Cellulose With Biological Acid Recycling (The Biosulfurol Process), Netherlands Organisation for Applied Scientific Research (TNO), Zeist, The Netherlands, 2007.

93. Y. H. P. Zhang, Z. Zhu, R. Rollin, and N. Sathitsuksanoh, "Advances in cellulose solvent- and organic solvent-based lignocellulose fractionation (COSLIF) in cellulose solvents," in Cellulose Solvents: For Analysis, Shaping and Chemical Modification, T. Liebert, T. Heinze, and K. Edgar, Eds., vol. 1033 of American Chemical Society Symposium Series, pp. 365–379, 2010.

94. J. Zhang, B. Zhang, J. Zhang, L. Lin, S. Liu, and P. Ouyang, "Effect of phosphoric acid pretreatment on enzymatic hydrolysis of microcrystalline cellulose," Biotechnology Advances, vol. 28, no. 5, pp. 613–619, 2010.

95. N. Sathitsuksanoh, Z. Zhu, and Y. H. P. Zhang, "Cellulose solvent- and organic solvent-based lignocellulose fractionation enabled efficient sugar release from a variety of lignocellulosic feedstocks," Bioresource Technology, vol. 117, pp. 228–233, 2012.

96. N. Sathitsuksanoh, Z. Zhu, T.-J. Ho, M.-D. Bai, and Y.-H. P. Zhang, "Bamboo saccharification through cellulose solvent-based biomass pretreatment followed by enzymatic hydrolysis at ultra-low cellulase loadings," Bioresource Technology, vol. 101, no. 13, pp. 4926–4929, 2010.

97. N. Sathitsuksanoh, Z. Zhu, and Y. H. P. Zhang, "Cellulose solvent-based pretreatment for corn stover and avicel: concentrated phosphoric acid versus ionic liquid [BMIM]Cl," Cellulose, vol. 19, no. 4, pp. 1161–1172, 2012.

98. J. Xu, X. Zhang, and J. J. Cheng, "Pretreatment of corn stover for sugar production with switchgrass-derived black liquor," Bioresource Technology, vol. 111, pp. 255–260, 2012.

99. J. E. Lindberg, I. E. Ternrud, and O. Theander, "Degradation rate and chemical composition of different types of Alkali-treated straws during rumen digestion," Journal of the Science of Food and Agriculture, vol. 35, pp. 500–506, 1984.

100. X. Zhao, L. Zhang, and D. Liu, "Pretreatment of Siam weed stem by several chemical methods for increasing the enzymatic digestibility," Biotechnology Journal, vol. 5, no. 5, pp. 493–504, 2010.

101. H. Y. Yoo, S. B. Kim, H. S. Choi, K. Kim, C. Park, and S. W. Kim, "Optimization of sodium hydroxide pretreatment of canola agricultural residues for fermentable sugar production using statistical method," in Proceedings of the International Conference on Future Environment and Energy (IPCBEE ‹12), vol. 28, 2012.

102. P. Boonsawang, Y. Subkaree, and T. Srinorakutara, "Ethanol production from palm pressed fiber by prehydrolysis prior to simultaneous saccharification and fermentation (SSF)," Biomass and Bioenergy, vol. 40, pp. 127–132, 2012.

103. Y.-S. Cheng, Y. Zheng, C. W. Yu, T. M. Dooley, B. M. Jenkins, and J. S. Vandergheynst, "Evaluation of high solids alkaline pretreatment of rice straw," Applied Biochemistry and Biotechnology, vol. 162, no. 6, pp. 1768–1784, 2010.

104. A. T. W. M. Hendriks and G. Zeeman, "Pretreatments to enhance the digestibility of lignocellulosic biomass," Bioresource Technology, vol. 100, no. 1, pp. 10–18, 2009.

105. F. Peng, J. Bian, J.-L. Ren, P. Peng, F. Xu, and R.-C. Sun, "Fractionation and characterization of alkali-extracted hemicelluloses from peashrub," Biomass and Bioenergy, vol. 39, pp. 20–30, 2012.

106. L. Yang, J. Cao, Y. Jin, et al., "Effects of sodium carbonate pretreatment on the chemical compositions and enzymatic saccharification of rice straw," Bioresource Technology, vol. 124, pp. 283–291, 2012.

107. L. Yang, J. Cao, J. Mao, and Y. Jin, "Sodium carbonate-sodium sulfite pretreatment for improving the enzymatic hydrolysis of rice straw," Industrial Crops and Products, vol. 43, pp. 711–717, 2013.

108. Y. Zhao, Y. Wang, J. Y. Zhu, A. Ragauskas, and Y. Deng, "Enhanced enzymatic hydrolysis of spruce by alkaline pretreatment at low temperature," Biotechnology and Bioengineering, vol. 99, no. 6, pp. 1320–1328, 2008.

109. R. A. Silverstein, Y. Chen, R. R. Sharma-Shivappa, M. D. Boyette, and J. Osborne, "A comparison of chemical pretreatment methods for improving saccharification of cotton stalks," Bioresource Technology, vol. 98, no. 16, pp. 3000–3011, 2007.

110. K. Mirahmadi, M. M. Kabir, A. Jeihanipour, K. Karimi, and M. J. Taherzadeh, "Alkaline pretreatment of spruce and birch to

improve bioethanol and biogas production," BioResources, vol. 5, no. 2, pp. 928–938, 2010.

111. X. Zhao, F. Peng, K. Cheng, and D. Liu, "Enhancement of the enzymatic digestibility of sugarcane bagasse by alkali-peracetic acid pretreatment," Enzyme and Microbial Technology, vol. 44, no. 1, pp. 17–23, 2009.

112. M. Ioelovich and E. Morag, "Study of enzymatic hydrolysis of mild pretreated lignocellulosic biomasses," BioResources, vol. 7, no. 1, pp. 1040–1052, 2012.

113. S. Ji and I. Lee, "Impact of cationic polyelectrolyte on the nanoshear hybrid alkaline pretreatment of corn stover: morphology and saccharification study," Bioresource Technology, vol. 133, pp. 45–50, 2013.

114. Y. Xing, L. X. Bu, K. Wang, and J. X. Jiang, "Pretreatment of furfural residues with alkaline peroxide to improve cellulose hydrolysis. Characterization of isolated lignin," Cellulose Chemistry and Technology, vol. 46, no. 3-4, pp. 249–260, 2012.

115. W. Cao, C. Sun, R. Liu, R. Yin, and X. Wu, "Comparison of the effects of five pretreatment methods on enhancing the enzymatic digestibility and ethanol production from sweet sorghum bagasse,"Bioresource Technology, vol. 111, pp. 215–221, 2012.

116. K. Minu, K. K. Jiby, and V. V. N. Kishore, "Isolation and purification of lignin and silica from the black liquor generated during the production of bioethanol from rice straw," Biomass and Bioenergy, vol. 39, pp. 210–217, 2012.

117. G. Banerjee, S. Car, J. S. Scott-Craig, D. B. Hodge, and J. D. Walton, "Alkaline peroxide pretreatment of corn stover: effects of biomass, peroxide, and enzyme loading and composition on yields of glucose and xylose," Biotechnology for Biofuels, vol. 4, article 16, 2011.

118. G. Banerjee, S. Car, J. S. Scott-Craig, M. S. Borrusch, and J. D. Walton, "Rapid optimization of enzyme mixtures for deconstruction of diverse pretreatment/biomass feedstock combinations," Biotechnology for Biofuels, vol. 3, article 22, 2010.

119. M. Falls and M. T. Holtzapple, "Oxidative lime pretreatment of alamo switchgrass," Applied Biochemistry and Biotechnology,

vol. 165, no. 2, pp. 506–522, 2011.
120. R. Kumar, G. Mago, V. Balan, and C. E. Wyman, "Physical and chemical characterizations of corn stover and poplar solids resulting from leading pretreatment technologies," Bioresource Technology, vol. 100, no. 17, pp. 3948–3962, 2009.
121. B. C. Saha and M. A. Cotta, "Lime pretreatment, enzymatic saccharification and fermentation of rice hulls to ethanol," Biomass and Bioenergy, vol. 32, no. 10, pp. 971–977, 2008.
122. R. Kumar and C. E. Wyman, "Cellulase adsorption and relationship to features of corn stover solids produced by leading pretreatments," Biotechnology and Bioengineering, vol. 103, no. 2, pp. 252–267, 2009.
123. L. L. G. Fuentes, S. C. Rabelo, R. M. Filho, and A. C. Costa, "Kinetics of lime pretreatment of sugarcane bagasse to enhance enzymatic hydrolysis," Applied Biochemistry and Biotechnology, vol. 163, no. 5, pp. 612–625, 2011.
124. W. E. Kaar and M. T. Holtzapple, "Using lime pretreatment to facilitate the enzymic hydrolysis of corn stover," Biomass and Bioenergy, vol. 18, no. 3, pp. 189–199, 2000.
125. R. J. Garlock, V. Balan, B. E. Dale et al., "Comparative material balances around pretreatment technologies for the conversion of switchgrass to soluble sugars," Bioresource Technology, vol. 102, no. 24, pp. 11063–11071, 2011.
126. S. C. Rabelo, R. M. Filho, and A. C. Costa, "A comparison between lime and alkaline hydrogen peroxide pretreatments of sugarcane bagasse for ethanol production," Applied Biochemistry and Biotechnology, vol. 148, no. 1–3, pp. 45–58, 2008.
127. R. Prado, X. Erdocia, L. Serrano, and J. Labldl, "Lignin purification with green solvents," Cellulose Chemistry and Technology, vol. 46, no. 3-4, pp. 221–225, 2012.
128. R. T. Elander, B. E. Dale, M. Holtzapple et al., "Summary of findings from the Biomass Refining Consortium for Applied Fundamentals and Innovation (CAFI): corn stover pretreatment," Cellulose, vol. 16, no. 4, pp. 649–659, 2009.
129. B. D. Schutt and M. A. Abraham, "Evaluation of a monolith reactor for the catalytic wet oxidation of cellulose," Chemical Engineering Journal, vol. 103, no. 1–3, pp. 77–88, 2004.

130. S. Rovio, A. Kallioinen, T. Tamminen, M. Hakola, M. Leskelä, and M. Siika-ahoa, "Catalysed alkaline oxidation as a wood fractionation technique," BioResources, vol. 7, no. 1, pp. 756–776, 2012.
131. H. Palonen, A. B. Thomsen, M. Tenkanen, A. S. Schmidt, and L. Viikari, "Evaluation of wet oxidation pretreatment for enzymatic hydrolysis of softwood," Applied Biochemistry and Biotechnology A, vol. 117, no. 1, pp. 1–17, 2004.
132. C. Martín, Y. González, T. Fernández, and A. B. Thomsen, "Investigation of cellulose convertibility and ethanolic fermentation of sugarcane bagasse pretreated by wet oxidation and steam explosion," Journal of Chemical Technology and Biotechnology, vol. 81, no. 10, pp. 1669–1677, 2006.
133. A. B. Thomsen and S. Schmidt, Further Development of Chemical and Biological Processes for Production of Bioethanol: Optimisation of Pre-Treatment Processes and Characterisation of Products, Risø-R-1110(EN), Risø National Laboratory, Roskilde, Denmark, 1999.
134. H. B. Klinke, L. Olsson, A. B. Thomsen, and B. K. Ahring, "Potential inhibitors from wet oxidation of wheat straw and their effect on ethanol production of Saccharomyces cerevisiae: wet oxidation and fermentation by yeast," Biotechnology and Bioengineering, vol. 81, no. 6, pp. 738–747, 2003.
135. E. Arvaniti, A. B. Bjerre, and J. E. Schmidt, "Wet oxidation pretreatment of rape straw for ethanol production," Biomass and Bioenergy, vol. 39, pp. 94–105, 2012.
136. S. Banerjee, R. Sen, R. A. Pandey et al., "Evaluation of wet air oxidation as a pretreatment strategy for bioethanol production from rice husk and process optimization," Biomass and Bioenergy, vol. 33, no. 12, pp. 1680–1686, 2009.
137. A. Petersson, M. H. Thomsen, H. Hauggaard-Nielsen, and A.-B. Thomsen, "Potential bioethanol and biogas production using lignocellulosic biomass from winter rye, oilseed rape and faba bean," Biomass and Bioenergy, vol. 31, no. 11-12, pp. 812–819, 2007.
138. A. B. Bjerre and A. S. Schmidt, Development of Chemical and Biological Processes for Production of Bioethanol: Optimization of the Wet Oxidation Process and Characterizaiton of Products,

Risø-R-967 (EN), Risø National Laboratory, Roskilde, Denmark, 1997.

139. A. O. Ayeni, S. Banerjee, J. A. Omoleye, et al., "Optimization of pretreatment conditions using full factorial design and enzymatic convertibility of shea tree sawdust," Biomass Bioenergy, vol. 48, pp. 130–138, 2013.

140. A. O. Ayeni, F. K. Hymore, S. N. Mudliar, et al., "Hydrogen peroxide and lime based oxidative pretreatment of wood waste to enhance enzymatic hydrolysis for a biorefinery: process parameters optimization using response surface methodology," Fuel, vol. 106, pp. 187–194, 2013.

141. M. G. Alriols, A. Tejado, M. Blanco, I. Mondragon, and J. Labidi, "Agricultural palm oil tree residues as raw material for cellulose, lignin and hemicelluloses production by ethylene glycol pulping process," Chemical Engineering Journal, vol. 148, no. 1, pp. 106–114, 2009.

142. H. L. Chum, L. J. Douglas, D. A. Feinberg, and H. A. Schroeder, Evaluation of Pretreatments of Biomass for Enzymatic Hydrolysis of Cellulose, US Department of Energy, Contract No. DE-AC02-83CHt0093, 1985.

143. B. W. Koo, B. C. Min, K. S. Gwak, et al., "Structural changes in lignin during organosolv pretreatment of Liriodendron tulipifera and the effect on enzymatic hydrolysis," Biomass Bioenergy, vol. 42, pp. 24–32, 2012.

144. N. Park, H.-Y. Kim, B.-W. Koo, H. Yeo, and I.-G. Choi, "Organosolv pretreatment with various catalysts for enhancing enzymatic hydrolysis of pitch pine (Pinus rigida)," Bioresource Technology, vol. 101, no. 18, pp. 7046–7053, 2010.

145. D. Haverty, K. Dussan, A. V. Piterina, J. J. Leahy, and M. H. B. Hayes, "Autothermal, single-stage, performic acid pretreatment of Miscanthus x giganteus for the rapid fractionation of its biomass components into a lignin/hemicellulose-rich liquor and a cellulase-digestible pulp," Bioresource Technology, vol. 109, pp. 173–177, 2012.

146. L. Mesa, E. González, C. Cara, M. González, E. Castro, and S. I. Mussatto, "The effect of organosolv pretreatment variables on enzymatic hydrolysis of sugarcane bagasse," Chemical Engineering Journal, vol. 168, no. 3, pp. 1157–1162, 2011.

147. X. Zhao and D. Liu, "Fractionating pretreatment of sugarcane bagasse by aqueous formic acid with direct recycle of spent liquor to increase cellulose digestibility-the Formiline process," Bioresource Technology, vol. 117, pp. 25–32, 2012.

148. J. J. Villaverde, P. Ligero, and A. de Vega, "Miscanthus x giganteus as a source of biobased products through organosolv fractionation: a mini review," The Open Agriculture Journal, vol. 4, pp. 102–110, 2010.

149. L. Kupiainen, J. Ahola, and J. Tanskanen, "Hydrolysis of organosolv wheat pulp in formic acid at high temperature for glucose production," Bioresource Technology, vol. 116, pp. 29–35, 2012.

150. Y. Kim, A. Yu, M. Han, G.-W. Choi, and B. Chung, "Enhanced enzymatic saccharification of barley straw pretreated by ethanosolv technology," Applied Biochemistry and Biotechnology, vol. 163, no. 1, pp. 143–152, 2011.

151. W. J. J. Huijgen, R. R. Van der Laan, and J. H. Reith, "Mordified organosolv as a fractionation process of lignocellulosic biomass for coproduction of fuels and chemicals," in Proceedings of the 16th Europeran Biomass Conference and Exhibition, Valencia, Spain, 2008.

152. E. K. Pye and J. H. Lora, "The AlcellTM process: a proven alternative to kraft pulping," Tappi Joumd, vol. 113, 1991.

153. C. Arato, E. K. Pye, and G. Gjennestad, "The lignol approach to biorefining of woody biomass to produce ethanol and chemicals," Applied Biochemistry and Biotechnology A, vol. 123, no. 1–3, pp. 871–882, 2005.

154. X. Pan, C. Arato, N. Gilkes et al., "Biorefining of softwoods using ethanol organosolv pulping: preliminary evaluation of process streams for manufacture of fuel-grade ethanol and co-products,"Biotechnology and Bioengineering, vol. 90, no. 4, pp. 473–481, 2005.

155. X. Pan, N. Gilkes, J. Kadla et al., "Bioconversion of hybrid poplar to ethanol and co-products using an organosolv fractionation process: optimization of process yields," Biotechnology and Bioengineering, vol. 94, no. 5, pp. 851–861, 2006.

156. A. Hideno, A. Kawashima, T. Endo, K. Honda, and M. Morita, "Ethanol-based organosolv treatment with trace hydrochloric

acid improves the enzymatic digestibility of Japanese cypress (Chamaecyparis obtusa) by exposing nanofibers on the surface," Bioresource Technology, vol. 132, pp. 64–70, 2013.
157. M. G. Alriols, A. García, R. Llano-ponte, and J. Labidi, "Combined organosolv and ultrafiltration lignocellulosic biorefinery process," Chemical Engineering Journal, vol. 157, no. 1, pp. 113–120, 2010.
158. X. Pan, J. F. Kadla, K. Ehara, N. Gilkes, and J. N. Saddler, "Organosolv ethanol lignin from hybrid poplar as a radical scavenger: relationship between lignin structure, extraction conditions, and antioxidant activity," Journal of Agricultural and Food Chemistry, vol. 54, no. 16, pp. 5806–5813, 2006.
159. H. Y. Kim, K. S. Gwak, S. Y. Lee, H. S. Jeong, K. O. Ryu, and I. G. Choi, "Biomass characteristics and ethanol production of yellow poplar (Liriodendron tulipifera) treated with slurry composting and biofiltration liquid as fertilizer," Biomass Bioenergy, vol. 42, pp. 10–17, 2012.
160. N. Brosse, R. E. L. Hage, P. Sannigrahi, and A. Ragauskas, "Dilute sulphuric acid and ethanol organosolv pretreatment of miscanthus x giganteus," Cellulose Chemistry and Technology, vol. 44, no. 1–3, pp. 71–78, 2010.
161. Z. Li, Z. Jiang, B. Fei, et al., "Ethanol organosolv pretreatment of bamboo for efficient enzymatic saccharification," BioResources, vol. 7, no. 3, pp. 3452–3462, 2012.
162. C. Wang, F. Zhou, Z. Yang et al., "Hydrolysis of cellulose into reducing sugar via hot-compressed ethanol/water mixture," Biomass and Bioenergy, vol. 42, pp. 143–150, 2012.
163. X. Zhao, K. Cheng, and D. Liu, "Organosolv pretreatment of lignocellulosic biomass for enzymatic hydrolysis," Applied Microbiology and Biotechnology, vol. 82, no. 5, pp. 815–827, 2009.
164. Y. N. Guragain, J. De Coninck, F. Husson, A. Durand, and S. K. Rakshit, "Comparison of some new pretreatment methods for second generation bioethanol production from wheat straw and water hyacinth," Bioresource Technology, vol. 102, no. 6, pp. 4416–4424, 2011.
165. S. Han, J. Li, S. Zhu et al., "Potential applications of ionic liquids

in wood related industries,"BioResources, vol. 4, no. 2, pp. 825–834, 2009.

166. A. P. Dadi, C. A. Schall, and S. Varanasi, "Mitigation of cellulose recalcitrance to enzymatic hydrolysis by ionic liquid pretreatment," Applied Biochemistry and Biotechnology, vol. 137–140, no. 1–12, pp. 407–421, 2007.

167. M. Mora-Pale, L. Meli, T. V. Doherty, R. J. Linhardt, and J. S. Dordick, "Room temperature ionic liquids as emerging solvents for the pretreatment of lignocellulosic biomass," Biotechnology and Bioengineering, vol. 108, no. 6, pp. 1229–1245, 2011.

168. T. Vancov, A.-S. Alston, T. Brown, and S. McIntosh, "Use of ionic liquids in converting lignocellulosic material to biofuels," Renewable Energy, vol. 45, pp. 1–6, 2012.

169. H. Wu, M. Mora-Pale, J. Miao, T. V. Doherty, R. J. Linhardt, and J. S. Dordick, "Facile pretreatment of lignocellulosic biomass at high loadings in room temperature ionic liquids," Biotechnology and Bioengineering, vol. 108, no. 12, pp. 2865–2875, 2011.

170. N. Muhammad, Z. Man, M. A. Bustam, M. I. A. Mutalib, C. D. Wilfred, and S. Rafiq, "Dissolution and delignification of bamboo biomass using amino acid-based ionic liquid," Applied Biochemistry and Biotechnology, vol. 165, no. 3-4, pp. 998–1009, 2011.

171. J. Holm and U. Lassi, Ionic Liquids in the Pretreatment of Lignocellulosic Biomass, Ionic Liquids: applications and perspectives, Edited by A. Kokorin, InTech, 2011.

172. R. Q. Xie, X. Y. Li, and Y. F. Zhang, "Cellulose pretreatment with 1-methyl-3-ethylimidazolium dimethylphosphate for enzymatic hydrolysis," Cellulose Chemistry and Technology, vol. 46, no. 5-6, pp. 349–356, 2012.

173. S. P. M. Ventura, L. D. F. Santos, J. A. Saraiva, and J. A. P. Coutinho, "Concentration effect of hydrophilic ionic liquids on the enzymatic activity of Candida antarctica lipase B.," World Journal of Microbiology and Biotechnology, vol. 28, pp. 2303–2310, 2012.

174. L. Li, S. T. Yu, F. S. Liu, C. S. Xie, and C. Z. Xu, "Efficient enzymatic in situ saccharificatio of cellulose in aqeous-ionic liquid media by microwave treatment," BioResources, vol. 6, no. 4, pp. 4494–

4504, 2011.
175. Z. Zhang, I. M. O›Hara, and W. O. S. Doherty, "Pretreatment of sugarcane bagasse by acid-catalysed process in aqueous ionic liquid solutions," Bioresource Technology, vol. 120, pp. 149–156, 2012.
176. Q. Wang, Y. Wu, and S. Zhu, "Use of ionic liquids for improvement of cellulosic ethanol production," BioResources, vol. 6, no. 1, pp. 1–2, 2011.
177. T. V. Doherty, M. Mora-Pale, S. E. Foley, R. J. Linhardt, and J. S. Dordick, "Ionic liquid solvent properties as predictors of lignocellulose pretreatment efficacy," Green Chemistry, vol. 12, no. 11, pp. 1967–1975, 2010.
178. A. S. Amarasekara and B. Wiredu, "Brönsted acidic ionic liquid 1-(1-propylsulfonic)-3-methylimidazolium-chloride catalyzed hydrolysisof D-cellobiose in aqueous medium," International Journal of Carbohydrate Chemistry, vol. 2012, Article ID 948652, 6 pages, 2012. ·
179. H. Xie, W. Liu, and Z. K. Zhao, "Lignocellulose pretreatment by ionic liquids: a promising start point for bio-energy production," Biomass Conversion, pp. 123–144, 2012.
180. G. Brodeur, E. Yau, B. Badal, and J. Collier, "Chemical and physicochemical pretreatment of lignocellulosic biomass: a review," Enzyme Research, vol. 2011, Article ID 787532, 17 pages, 2011. ·
181. X.-F. Tian, Z. Fang, D. Jiang, and X.-Y. Sun, "Pretreatment of microcrystalline cellulose in organic electrolyte solutions for enzymatic hydrolysis," Biotechnology for Biofuels, vol. 4, article 53, 2011.
182. S. Singh, B. A. Simmons, and K. P. Vogel, "Visualization of biomass solubilization and cellulose regeneration during ionic liquid pretreatment of switchgrass," Biotechnology and Bioengineering, vol. 104, no. 1, pp. 68–75, 2009.
183. G. Papa, P. Varanasi, L. Sun, et al., "Exploring the effect of different plant lignin content and composition on ionic liquid pretreatment efficiency and enzymatic saccharification of Eucalyptus globulus L. Mutants," Bioresource Technology, vol. 117, pp. 352–359, 2012.
184. B. Li, J. Asikkala, I. Filpponen, and D. S. Argyropoulos, "Factors

affecting wood dissolution and regeneration of ionic liquids," Industrial and Engineering Chemistry Research, vol. 49, no. 5, pp. 2477–2484, 2010.

185. A. P. Dadi, S. Varanasi, and C. A. Schall, "Enhancement of cellulose saccharification kinetics using an ionic liquid pretreatment step," Biotechnology and Bioengineering, vol. 95, no. 5, pp. 904–910, 2006.

186. M. Ungurean, F. Fițigău, C. Paul, A. Ursoiu, and F. Peter, "Ionic liquid pretreatment and enzymatic hydrolysis of wood biomass," World Academy of Science, Engineering and Technology, vol. 76, pp. 387–391, 2011.

187. S. Morales-delaRosa, J. M. Campos-Martin, and J. L. G. Fierro, "High glucose yields from the hydrolysis of cellulose dissolved in ionic liquids," Chemical Engineering Journal, vol. 181-182, pp. 538–541, 2012.

188. P. Weerachanchai, S. S. J. Leong, M. W. Chang, C. B. Ching, and J.-M. Lee, "Improvement of biomass properties by pretreatment with ionic liquids for bioconversion process," Bioresource Technology, vol. 111, pp. 453–459, 2012.

189. R. Arora, C. Manisseri, C. Li et al., "Monitoring and analyzing process streams towards understanding ionic liquid pretreatment of switchgrass (Panicum virgatum L.)," Bioenergy Research, vol. 3, no. 2, pp. 134–145, 2010.

190. J. A. Perez-Pimienta, M. G. Lopez-Ortega, P. Varanasi, et al., "Comparison of the impact of ionic liquid pretreatment on recalcitrance of agave bagasse and switchgrass," Bioresource Technology, vol. 127, pp. 18–24, 2013.

191. Z. Zhi-Guo and C. Hong-Zhang, "Enhancement of the enzymatic hydrolysis of wheat straw by pretreatment with 1-allyl-3-methylimidazolium chloride ([Amim]Cl)," African Journal of Biotechnology, vol. 11, no. 31, pp. 8032–8037, 2012.

192. M. Lucas, G. L. Wagner, Y. Nishiyama et al., "Reversible swelling of the cell wall of poplar biomass by ionic liquid at room temperature," Bioresource Technology, vol. 102, no. 6, pp. 4518–4523, 2011.

193. G. Cheng, P. Varanasi, C. Li et al., "Transition of cellulose crystalline structure and surface morphology of biomass as a

function of ionic liquid pretreatment and its relation to enzymatic hydrolysis," Biomacromolecules, vol. 12, no. 4, pp. 933–941, 2011.

194. X.-D. Hou, T. J. Smith, N. Li, and M.-H. Zong, "Novel renewable ionic liquids as highly effective solvents for pretreatment of rice straw biomass by selective removal of lignin," Biotechnology and Bioengineering, vol. 109, no. 10, pp. 2484–2493, 2012.

195. D. Yuan, K. Rao, S. Varanasi, and R. Relue, "A viable method and configuration for fermenting biomass sugars to ethanol using native Saccharomyces cerevisiae," Bioresource Technology, vol. 117, pp. 92–98, 2012.

196. Z. Qiu, G. M. Aita, and M. S. Walker, "Effect of ionic liquid pretreatment on the chemical composition, structure and enzymatic hydrolysis of energy cane bagasse," Bioresource Technology, vol. 117, pp. 251–256, 2012.

197. T. N. Ang, G. C. Ngoh, A. S. M. Chua, and M. G. Lee, "Elucidation of the effect of ionic liquid pretreatment on rice husk via structural analyses," Biotechnology for Biofuels, vol. 5, article 67, 2012.

198. K. Ninomiya, T. Yamauchi, M. Kobayashi, et al., "Cholinium carboxylate ionic liquids for pretreatment of lignocellulosic materials to enhance subsequent enzymatic saccharification,"Biochemical Engineering Journal, vol. 71, pp. 25–29, 2013.

199. E. Gnansounou, "Production and use of lignocellulosic bioethanol in Europe: current situation and perspectives," Bioresource Technology, vol. 101, no. 13, pp. 4842–4850, 2010.

200. E. Tomás-Pejó, P. Alvira, M. Ballesteros, and M. J. Negro, "Pretreatment technologies for lignocellulose-to-bioethanol conversion," in Biofuels Alternative Feedstocks and Conversion Processes, A. Pandey, C. Larroche, S. C. Ricke, C. G. Dussap, and E. Gnansounou, Eds., chapter 7, p. 170, Elsevier, 2011.

201. G. Rødsrud, M. Lersch, and A. Sjöde, "History and future of world's most advanced biorefinery in operation," Biomass Bioenergy, vol. 46, pp. 46–59, 2012.

202. I. M. O›Hara, L. A. Edye, W. O. S. Doherty, and G. Kent, Demonstration of Cellulosic Ethanol Production From Sugarcane Bagasse in Australia: The Mackay Renewable Biocommodities

Pilot Plant, International Society of Sugar Cane Technologists, Veracruz, Mexico, 2010.
203. X. Q. Zhao, L. H. Zi, F. W. Bai, H. L. Lin, X. M. Hao, and G. J. Yue, "Bioethanol from lignocellulosic biomass," Advances in Biochemical Engineering/Biotechnology, vol. 128, pp. 25–51, 2012.
204. S. Hyvärinen, P. Virtanen, D. Y. Murzin, and J.-P. Mikkola, "Towards ionic liquid fractionation of lignocellulosics for fermentable sugars," Cellulose Chemistry and Technology, vol. 44, no. 4-6, pp. 187–195, 2010.
205. J. I. Park, E. J. Steen, H. Burd, et al., "A thermophilic ionic liquid-tolerant cellulase cocktail for the production of cellulosic biofuels," PLoS ONE, vol. 7, no. 5, article e37010, 2012.
206. H. Guo, X. Qi, L. Li, and R. L. Smith Jr., "Hydrolysis of cellulose over functionalized glucose-derived carbon catalyst in ionic liquid," Bioresource Technology, vol. 116, pp. 355–359, 2012.
207. S.-J. Kim, A. A. Dwiatmoko, J. W. Choi, Y.-W. Suh, D. J. Suh, and M. Oh, "Cellulose pretreatment with 1-n-butyl-3-methylimidazolium chloride for solid acid-catalyzed hydrolysis," Bioresource Technology, vol. 101, no. 21, pp. 8273–8279, 2010.
208. Z. Zhu, M. Zhu, and Z. Wu, "Pretreatment of sugarcane bagasse with NH_4OH-H_2O_2 and ionic liquid for efficient hydrolysis and bioethanol production," Bioresource Technology, vol. 119, pp. 199–207, 2012.
209. G. Bokinsky, P. P. Peralta-Yahya, A. George et al., "Synthesis of three advanced biofuels from ionic liquid-pretreated switchgrass using engineered Escherichia coli," Proceedings of the National Academy of Sciences of the United States of America, vol. 108, no. 50, pp. 19949–19954, 2011.
210. M. T. García-Cubero, G. González-Benito, I. Indacoechea, M. Coca, and S. Bolado, "Effect of ozonolysis pretreatment on enzymatic digestibility of wheat and rye straw," Bioresource Technology, vol. 100, no. 4, pp. 1608–1613, 2009.
211. N. Schultz-Jensen, Z. Kádár, A. B. Thomsen, H. Bindslev, and F. Leipold, "Plasma-assisted pretreatment of wheat straw for ethanol production," Applied Biochemistry and Biotechnology, vol. 165, no. 3-4, pp. 1010–1023, 2011.

212. R. Travaini, M. D. Otero, M. Coca, R. Da-Silva, and S. Bolado, "Sugarcane bagasse ozonolysis pretreatment: effect on enzymatic digestibility and inhibitory compound formation," Bioresource Technology, vol. 133, pp. 332–339, 2013.
213. P. Sannigrahi, F. Hu, Y. Pu, and A. Ragauskasa, "Novel oxidative pretreatment of Loblolly Pine, Sweetgum, and Miscanthus by ozone," Journal of Wood Chemistry and Technology, vol. 32, pp. 361–375, 2012.
214. N. Schultz-Jensen, F. Leipold, H. Bindslev, and A. B. Thomsen, "Plasma-assisted pretreatment of wheat straw," Applied Biochemistry and Biotechnology, vol. 163, no. 4, pp. 558–572, 2011.
215. D. Rodriguez-Gomez, L. Lehmann, N. Schultz-Jensen, A. B. Bjerre, and T. J. Hobley, "Examining the potential of plasma-assisted pretreated wheat straw for enzyme production by Trichoderma reesei,"Applied Biochemistry and Biotechnology, vol. 166, pp. 2051–2063, 2012.
216. S. K. Uppal, R. Kaur, and P. Sharma, "Optimization of chemical pretreatment and acid saccharification for conversion of sugarcane bagasse to ethanol," Sugar Tech, vol. 13, no. 3, pp. 214–219, 2011.
217. D. Mishima, M. Tateda, M. Ike, and M. Fujita, "Comparative study on chemical pretreatments to accelerate enzymatic hydrolysis of aquatic macrophyte biomass used in water purification processes,"Bioresource Technology, vol. 97, no. 16, pp. 2166–2172, 2006.
218. M. Lucas, S. K. Hanson, G. L. Wagner, D. B. Kimball, and K. D. Rector, "Evidence for room temperature delignification of wood using hydrogen peroxide and manganese acetate as a catalyst,"Bioresource Technology, vol. 119, pp. 174–180, 2012.
219. G. Banerjee, S. Car, T. Liu et al., "Scale-up and integration of alkaline hydrogen peroxide pretreatment, enzymatic hydrolysis, and ethanolic fermentation," Biotechnology and Bioengineering, vol. 109, no. 4, pp. 922–931, 2012.
220. R.-S. Yao, H.-J. Hu, S.-S. Deng, H. Wang, and H.-X. Zhu, "Structure and saccharification of rice straw pretreated with sulfur trioxide micro-thermal explosion collaborative dilutes alkali," Bioresource Technology, vol. 102, no. 10, pp. 6340–6343, 2011.

221. F. Li, R. Yao, H. Wang, H. Hu, and R. Zhang, "Process optimization for sugars production from rice straw via pretreatment with sulphur trioxide micro-thermal explosion," BioResources, vol. 7, no. 3, pp. 3355–3366, 2012.
222. P. Sannigrahi, A. J. Ragauskas, and S. J. Miller, "Chlorine dioxide treatment of biomass feedstock," U.S. Patent Application Publishing, Patent No. US 8, 497, 097 B2, 2012.
223. L. Bu, Y. Xing, H. Yu, Y. Gao, and J. Jiang, "Comparative study of sulfite pretreatments for robust enzymatic saccharification of corn cob residue," Biotechnology for Biofuels, vol. 5, article 8, 2012.
224. Y. Jin, L. Yang, H. Jameel, H. M. Chang, and R. Phillips, "Sodium sulfite-formaldehyde pretreatment of mixed hardwoods and its effect on enzymatic hydrolysis," Bioresource Technology, vol. 135, pp. 109–115, 2013.
225. Q. Li, Y. Gao, H. Wang, and B. Li, "Comparison of different alkali-based pretreatments of corn stover for improving enzymatic saccharification," Bioresource Technology, vol. 125, pp. 193–199, 2012.
226. K.-K. Cheng, W. Wang, J.-A. Zhang, Q. Zhao, J.-P. Li, and J.-W. Xue, "Statistical optimization of sulfite pretreatment of corncob residues for high concentration ethanol production," Bioresource Technology, vol. 102, no. 3, pp. 3014–3019, 2011.
227. J. Y. Zhu, W. Zhu, P. Obryan et al., "Ethanol production from SPORL-pretreated lodgepole pine: preliminary evaluation of mass balance and process energy efficiency," Applied Microbiology and Biotechnology, vol. 86, no. 5, pp. 1355–1365, 2010.
228. D. S. Zhang, Q. Yang, J. Y. Zhu, and X. J. Pan, "Sulfite, (SPORL) pretreatment of switchgrass for enzymatic saccharification," Bioresource Technology, vol. 129, pp. 127–134, 2013.
229. L. Shuai, Q. Yang, J. Y. Zhu et al., "Comparative study of SPORL and dilute-acid pretreatments of spruce for cellulosic ethanol production," Bioresource Technology, vol. 101, no. 9, pp. 3106–3114, 2010.
230. Q. Yang and X. Pan, "Pretreatment of Agave americana stalk for enzymatic saccharification," Bioresource Technology, vol. 126,

pp. 336–340, 2012.

231. H. Liu and J. Y. Zhu, "Eliminating inhibition of enzymatic hydrolysis by lignosulfonate in unwashed sulfite-pretreated aspen using metal salts," Bioresource Technology, vol. 101, no. 23, pp. 9120–9127, 2010.

232. G. P. da Silva, M. Mack, and J. Contiero, "Glycerol: a promising and abundant carbon source for industrial microbiology," Biotechnology Advances, vol. 27, no. 1, pp. 30–39, 2009.

233. F. Yang, M. Hanna, and R. Sun, "Value-added uses for crude glycerol—a byproduct of biodiesel production," Biotechnology for Biofuels, vol. 5, article 13, 2012.

234. H. Zhao, J. H. Kwak, Y. Wang, J. A. Franz, J. M. White, and J. E. Holladay, "Interactions between cellulose and N-methylmorpholine-N-oxide," Carbohydrate Polymers, vol. 67, no. 1, pp. 97–103, 2007.

235. Q. Li, G.-S. Ji, Y.-B. Tang, X.-D. Gu, J.-J. Fei, and H.-Q. Jiang, "Ultrasound-assisted compatible in situ hydrolysis of sugarcane bagasse in cellulase-aqueous-N-methylmorpholine-N-oxide system for improved saccharification," Bioresource Technology, vol. 107, pp. 251–257, 2012.

236. M. Shafiei, K. Karimi, and M. J. Taherzadeh, "Pretreatment of spruce and oak by N-methylmorpholine-N-oxide (NMMO) for efficient conversion of their cellulose to ethanol,"Bioresource Technology, vol. 101, no. 13, pp. 4914–4918, 2010.

237. A. Goshadrou, K. Karimi, and M. J. Taherzadeh, "Ethanol and biogas production from birch by NMMO pretreatment," Biomass Bioenergy, vol. 49, pp. 95–101, 2013.

238. N. Poornejad, K. Karimi, and T. Behzad, "Improvement of saccharification and ethanol production from rice straw by NMMO and [BMIM][OAc] pretreatments," Industrial Crops and Products, vol. 41, pp. 408–413, 2013.

239. P. R. Lennartsson, C. Niklasson, and M. J. Taherzadeh, "A pilot study on lignocelluloses to ethanol and fish feed using NMMO pretreatment and cultivation with zygomycetes in an air-lift reactor,"Bioresource Technology, vol. 102, no. 6, pp. 4425–4432, 2011.

240. M. Shafiei, K. Karimi, and M. J. Taherzadeh, "Techno-economical study of ethanol and biogas from spruce wood by NMMO-pretreatment and rapid fermentation and digestion," Bioresource Technology, vol. 102, no. 17, pp. 7879–7886, 2011.
241. J. C. López-Linares, I. Romero, M. Moya, et al., "Pretreatment of olive tree biomass with $FeCl_3$ prior enzymatic hydrolysis," Bioresource Technology, vol. 128, pp. 180–187, 2013.
242. H. Li and J. Xu, "Optimization of microwave-assisted calcium chloride pretreatment of corn stover,"Bioresource Technology, vol. 127, pp. 112–118, 2013.

Chapter 9

Evaluation of Methane Yield on Mesophilic-Dry Anaerobic Digestion of Piggery Manure Mixed with Chaff for Agricultural Area

Dong-Heui Kwak[1], Mi-Sug Kim[1], Jae-Seung Kim[2], Young-Youl Oh[3], Soon-Ok Noh[4,5], Byung-Ok So[4,5], Su-Young Jung[4,5], Su-Jin Jung[4,5], and Soo-Wan Chae[4,5]

[1]Department of Environmental and Chemical Engineering, Seonam University, Namwon, Republic of Korea

[2]Department of Environmental Engineering, Chonbuk National University, Jeonju, Republic of Korea

[3]Yoyo Korea Agricultural Association, Jeongehb, Republic of Korea

[4]Clinical Trial Center for Functional Foods, Chonbuk National University Hospital, Jeonju, Republic of Korea

[5]Department of Medical Nutrition Therapy, Chonbuk National University, Medical School, Jeonju, Republic of Korea

ABSTRACT

A mesophilic-dry anaerobic digestion process is valid in treating high-concentration substrates containing low moisture content. It has merits of lower wastewater discharge and lower heat capacity required in maintaining reactor temperature as compared with a thermophilic-wet anaerobic digestion process. In fact, chaff can be easily obtained in farming areas and used as a mixture substrate as one of bulking agents for controlling moisture and supplying carbon. For this reason, this study applies the chaff to improve livestock manure, which contains high moisture content and is discharged from domestic pig farms. This study aims at verifying its feasibility for improving methane production efficiency on a basis of BMP (Biochemical Methane Potential) assay obtained through a series of experiments. Finding results were methane gas production and gas production per volatile solid (VS) added, and methane gas production among biogas production was increased as the chaff added in the piggery manure was increased. According to experimental results for improving the methane production efficiency, mixture of the chaff and the piggery manure played an important role in controlling the moisture content and improving the methane gas production rate, and also verified its feasibility in the mesophilic-dry anaerobic digestion process indicating relatively less difficulty for operation and management.

INTRODUCTION

A methane fermentation process has an advantage in treating organic contaminants for preventing environmental pollution when comparing a conventional aerobic treatment process. Naturally, the methane fermentation process has a combination limit of processes but it is relatively useful in aspects of energy production and resource collection. The methane fermentation process such as an anaerobic digestion process is a skill studied and used for a long time and is recently being magnified in a situation as an international concern focusing on climate change control and renewable energy demand. Especially, a biogas plant, one of the methane fermentation skills, has been used in many countries and known as one of effective strategy techniques for bio-fuel production [1].

Domestic livestock manure emission classified by livestock types has a component ratio as follows; 57.6% piggery manure (740,000 m^3/d) and 42.4% cow manure (540,000 m^3/d) [2], and dairy cow manure of the cow manure is emitted in the overcrowded area such as farms but few Korean native cattle are raised in small farmers and there are many bad cases in collecting and in treating Korean native cattle manure. The organic content is a raw matter to produce the methane and exists in the piggery manure as low as 2% to 5%. Thus, utilization of the piggery manure as a substrate is low and also it is known well that a stable operation of an anaerobic digestion tank is difficult because a fluctuation range of the organic content is large periodically and the moisture content is high enough [3].

The anaerobic digestion process is divided into a wet process and a dry process according to solid content or moisture content of the substrate used. Until the mid-1980s, the wet process has been mainly applied in the field using waste matters within 10% solid content as the substrate. With EU as the center rapidly from the 1990s, however, the dry process has been developed to digest organic waste matters containing the solid content over 20% [4]. In treating the high-concentration substrate having low moisture content, the dry process requires low heat capacity to maintain the reactor temperature and discharges low wastewater after treatment [5]. However, it is not valid to put the livestock manure into the dry process directly because the livestock manure emitted from domestic piggery farms contains a great deal of moisture and the solid content of the livestock manure is very low. Meanwhile, the bulking agents such as rice straws, chaff, dead leaves fragments, sawdust, etc. are easy to obtain in the farm area and such agricultural byproducts have been used as the bulking agent for composing manure from old times and also as the carbon supplement for maintaining the proper C/N ratio. Practically, most of domestic livestock farms are located in the farming settlement that is producing a great deal of the bulking agent. In adopting the biogas plant in the domestic farming areas, the dry anaerobic digestion process is in a more advantageous situation than the wet process when considering realistic conditions, In Europe recently, studies on the dry digestion operation for the municipal organic solid waste are actively proceeding to reduce waste amounts for landfill and to produce the bio-energy [6,7]. However, previous studies mainly present that the operation results for high temperature (50°C - 60°C) conditions and continuous operation cases are also pretty rare [8].

To analyze the ultimate methane production rate (mL/gVS$_{added}$) caused by organic matters as the substrate for the anaerobic digestion process, it measures the methane amounts produced during the anaerobic batch incubation period and cumulative methane formula can be used to determine the methane production yield based on the observed data. Representative models such as Modified Gompertz model or Exponential model are used to analyze experiment data obtained through the methane production potential test [9, 10]. Using those models described in Equations (1) and (2) as below, comparative studies are variously proceeding to determine the ultimate methane production yield of the substrates related to diverse components [11, 12].

Modified Gompertz Model Equation [13]:

$$M = M_o \times \exp\left\{-\exp\left[\frac{R_m \times e}{M_o}(\lambda - t) + 1\right]\right\} \quad (1)$$

Where, M: cumulative methane production yield (mLCH$_4$/g-VS)
t: incubation time of an anaerobic digestion tank (days)
M_o: ultimate methane production yield (mL-CH$_4$/g-VS)
R_m: maximum methane production rate (mL-CH$_4$/gVS·day)
e: exp (1) = 2.71828182
λ: lag phase, days Exponential Model Equation:

$$B = B_o\left(1 - e^{-kt}\right) \quad (2)$$

Where, B: cumulative methane production yield (mLCH$_4$/g-VS)
t: incubation time of an anaerobic digestion tank (days)
B_o: ultimate methane production yield (mL-CH$_4$/g-VS)
k: 1st order reaction rate constant (day^{-1}).

With the purpose of energy resource recovery through the methane production due to the piggery manure in the farming area, this study conducts a series of experiments with the dry anaerobic digestion using the chaff. The chaff is easily obtained in the farming area as one of the byproducts and as the substrate to mix with the manure as well as to control the moisture. The anaerobic digestion has been conducted in the single-phase mesophilic condition, BMP (biochemical methane potential) assay has been applied to estimate the methane production potential due to the several mixture ratios between the piggery manure and the chaff based on the experimental results, and this study has been accomplished to find a way to improve the methane production efficiency.

MATERIALS AND METHODS

Experimental Equipments and Operation Conditions

In experiments, a batchwise reactor of methane yield is prepared for single-phase anaerobic digestion as shown in Figure 1. For a dry methane production process of livestock manure as a main substrate, typical experimental conditions were adopted to examine gas production yield and responses characteristics. Experiments were set up to control pH if necessary and to incubate for 40 days as controlling to keep typical temperature for mesophilic digestion in a range of 35°C ± 1°C [14], and an additional agitator was excluded in the experiments.

Figure 1 describes a schematic diagram of a batchwise single phase digester for methane yield. 0.5 L serum bottles were set up to an incubator at a constant temperature. Operation conditions were monitored for every serum bottle in different mix proportions between piggery manure as a main substrate and chaff as a mixture substrate to control the moisture. Major items such as pH change and the gas production yield from the serum bottle were monitored due to each condition. The operator also measured generating-capacity and methane content in gas collected in a teflon bag through an exhaust pipe in the middle of a gas-tight rubber stopper of the serum bottle at a constant time interval, every 3 days. Additionally, taking small amounts from all of samples and putting them into each 50 mL-tube, the operator measured its weight per VS (volatile solid) changed with operating under the same conditions.

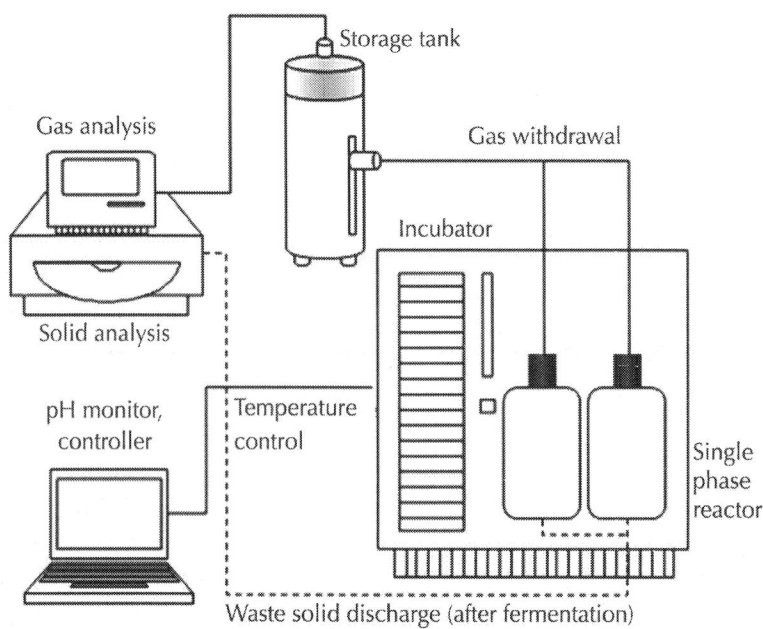

Figure 1: Schematic diagram of batchwise single phase digester for methane yield.

Table 1 presents operation conditions and a substrate composition in the single phase digester. For two operation parameters such as the substrate concentration and the solid content, the methane production was estimated in different mixing ratios between the piggery manure and the chaff. All samples except a control group (marked as Run 1) were added trace elements in order to minimize unstable effects of microbial growth due to lack of essential elements in the anaerobic digester.

Sample and Analysis Method

Livestock manure has been concerned in domestic water management as a non-point source. Among the livestock manure, piggery manure is newly applied as a main substrate for methane production. To improve the methane production, chaff in powder form (below than 100 mesh) is also mixed with the piggery manure to control the moisture as well as to improve the methane production. Major properties of the piggery manure are shown in Table 2 and chemical properties of all samples (Run 1 - Run 5) are described in Table 3. Also, Table 4 presents trace elements (mineral salts and trace metals) added for the safe operating condition. Water quality and solid matters were analyzed by the standard method (2005) in this study.

The piggery manure samples were collected in the retaining tank for gathering and mixing the manure of the pigsty before inputting washing water of a pig farm.

Table 1: Operation condition and substrate composition of single phase digester

Samples	Mixing ratio between manure and chaff	Substrate and solid content			
		Piggery manure (g)	Powder of chaff* (g/L)	Solid content* (%)	Trace elements
Run 1	2.50:1 (control)	300	120	35.4	non-spiked
Run 2	4.17:1	300	72	27.2	spiked
Run 3	3.13:1	300	95	30.6	spiked
Run 4	2.78:1	300	108	33.1	spiked
Run 5	2.50:1	300	120	35.7	spiked

Note: *Converted value into concentration.

Table 2: Chemical composition of piggery wastewater used as main substrate in this study

Description	Unit	Measured values
pH		8.7
BOD	mg/L	2205
COD_{cr}	mg/L	2221
$SCOD_{cr}$	mg/L	1126
T-N	mg/L	3439

T-P	mg/L	121.8
NH_3-N	mg/L	1304
P_4^{3-}-P	mg/L	4.3
Alkalinity	mg/L as $CaCO_3$	5800
Fixed solid	mg/L (%)	5900 (0.59)
Volatile solid	mg/L (%)	3300 (0.33)

Table 3: Chemical composition of experimental substrates (mixture of piggery manure and chaff)

Description	Unit	Run 1	Run 2	Run 3	Run 4	Run 5
pH		8.1	7.9	7.8	8.0	8.0
COD_{cr}	mg/kg	16,587	10,092	14,309	15,447	16,760
T-N	mg/kg	1583	3286	2988	1848	1643
T-P	mg/kg	151.4	189.2	169.2	159.2	153.2
NH_3-N	mg/kg	1122.6	1553.1	1416.0	1401.2	1204.8
Fixed solid	%	4.071	6.129	5.707	5.147	4.571
Volatile solid	%	3.048	2.984	2.879	2.966	3.071

Characteristics of target wastewater compiled from the piggery farm were indicated high pH around 8.7 and high alkalinity about 5800 mg/L as $CaCO_3$, which were typical in the livestock wastewater. The wastewater was measured in BOD (2205 mg/L) and $TCOD_{Cr}$ (2221 mg/L) and contained soluble COD_{Cr} ratio around 50.7%. Total solid concentration was 0.92% with 99.1% water content and the solid consisted of VS (35.9%) and inorganic solid (64.1%). Fraction of nutrients compared with the organic was as follows; $COD:NH_3$-$N:PO_4^{3-}$-$P = 51.7:30.3:1.0$.

Table 3 describes chemical compositions of experimental substrates in five different sample groups from Run 1 to Run 5 classified by different mixing ratios between the piggery manure and the chaff at identical conditions. The solid content was ranged from 27.2% to 35.4% and COD_{cr} was distributed from 10.092 mg/L to 16,760 mg/L. Ratios of the organic matter and nutrients (COD:T-N:T-P) for the sample groups (Run 1 - Run 5) were 109.6:10.5:1, 53.3:17.4:1, 84.6: 17.7:1, 97.0:11.6:1, and 109.4:10.7:1, respectively.

BMP Assay

A BMP (biochemical methane potential) assay was developed by Owen et al. [10] to evaluate potential efficiency for biodegradability of target livestock manure in an anaerobic process and it was analyzing organic concentration converted into CH_4. In this study, a serum bottle was filled with a target sample, covered with a butyl rubber septum, sealed with a reinforced plastic lid, and kept in an incubator at a constant temperature 35°C to induce anaerobic degradation. Gas producing capacity and its composition were analyzed every time interval and then a methane production rate was recorded. The control group (Run 1) was examined under the same conditions to modify other gas capacity and other effects occurred from the target experimental samples. Unlike the target experimental groups, the control group did not add trace elements consisting of mineral salts and trace metals. The gas producing capacity was measured using a glass syringe in the constant time interval and the methane production rate was analyzed using Gas Chromatography-Mass Spectrometer, GC-MS (6890 N Network GC system).

Table 4: Composition of mineral salts and trace metals

Concentration of mineral salts (mg/L)		Concentration of trace metals (mg/L)			
NH_4Cl	0.53	$MnCl_2 \cdot 4H_2O$	0.0005	$NaMoO_4 \cdot 2H_2O$	0.00001
$CaCl_2 \cdot 2H_2O$	0.075	H_3BO_3	0.00005	$CoCl_2 \cdot 6H_2O$	0.0005
$MgCl_2 \cdot 6H_2O$	0.1	$ZnCl_2$	0.00005	$NiCl_2 \cdot 6H_2O$	0.00005
$FeCl_2 \cdot 4H_2O$	0.02	$CuCl_2$	0.00003	Na_2SeO_3	0.00005

RESULT AND DISCUSSION

Experimental Result of Dry Mix Digestion

Piggery manure has had a difficult time in proceeding wet anaerobic digestion because of its high moisture content. For improving the anaerobic digestion of the piggery manure, it may be mixed with chaff obtained easily in farming areas. Table 5" target="_self"> Table 5 presents main experimental results of gas production and methane yields through 40 days digestion operating period for five sample groups.

In anaerobic degradation reactions of organic matters, substrate concentrations and physical and chemical compositions can effect on a reaction rate of hydrolysis and acid formation. Hence, this experiment set an equal amount of the livestock manure for all of samples to prevent the methane production rate from reducing.

Table 5: Experiment results of methane yields for five types of mixed substrates

Descriptions	Samples				
	Run 1	Run 2	Run 3	Run 4	Run 5
Cumulative volume of biogas yield (L)	3.21	3.48	3.69	4.19	4.76
Average daily biogas yield (L)	0.708	0.780	0.825	0.933	1.040
Maximum daily biogas yield (L)	1.070	1.160	1.230	1.397	1.587
Average methane fraction in biogas (%)	0.639	0.680	0.692	0.689	0.720
Maximum methane fraction in biogas (%)	0.735	0.782	0.791	0.842	0.849
Volatile solid (added) (g)	12.8	11.1	11.4	12.1	12.9
VS removal fraction (-)	0.667	0.843	0.808	0.833	0.797
Total biogas production rate (L/g-$VS_{removed}$)	0.376	0.372	0.396	0.416	0.463
Average daily biogas production rate (L/g-$VS_{removed}$·day)	0.102	0.109	0.121	0.116	0.139
Maximum daily biogas production rate (L/g-$VS_{removed}$·day)	0.198	0.201	0.208	0.216	0.235
Terminate duration of methane yield (days)	37 - 40	34 - 37	37 - 40	37 - 40	37 - 40

For considering the biogas production as shown in Table 5, total gas production and the gas production yield per VS added were also increased when the more chaff was mixed with the piggery manure. As mentioned above, the methane gas fraction in the biogas swung upward as the more chaff was added in the mixture. The addition of the chaff obviously improved the methane production rate and reduced the moisture content. In Figure 5, Run 1 and Run 2 used the constant mix proportion of the chaff but Run 1 without the trace elements produced less gas amounts compared to Run 2.

VS removal efficiency was estimated depending on measurement results of VS change amounts and was distinct from the change of the gas production yield. Although the VS were measured for the same sample under the identical conditions, in fact, there was a limitation in conforming tiny change of weight when the 0.5 L-serum bottle was measured directly. Thus, it may be possible to have some error because this study has measured a 0.05 L-additional container at the constant time interval.

Figure 2 indicates cumulative gas production rates of five sample groups for the digestion period. A large amount of gas was produced when the large amount of the chaff was added. Also, the daily gas production rate was slightly rapid at the sample group with the large amount of the chaff as shown in Figure 3 and a maximum gas production peak was appeared at 10 - 15 days after the digestion was started. Therefore, the sample group with the larger chaff content indicated the more gas production yield and the faster gas production rate, and the earlier maximum gas production peak.

Ultimate Methane Productivity Evaluation

Theoretical total methane productivities of piggery manure (516 L/g-VS by the hog and 530 L/g-VS by the sow) were higher than those of the cow manure (469 L/g-VS). The ultimate methane yield was in order as follows; 356 L/g-VS by the hog, 275 L/g-VS by the sow, and 148 L/g-VS by the milk cow. Also, the straw and the chaff were known to have much more methane yields as compared with the animal manure [15].

234 Direct Methane to Methanol

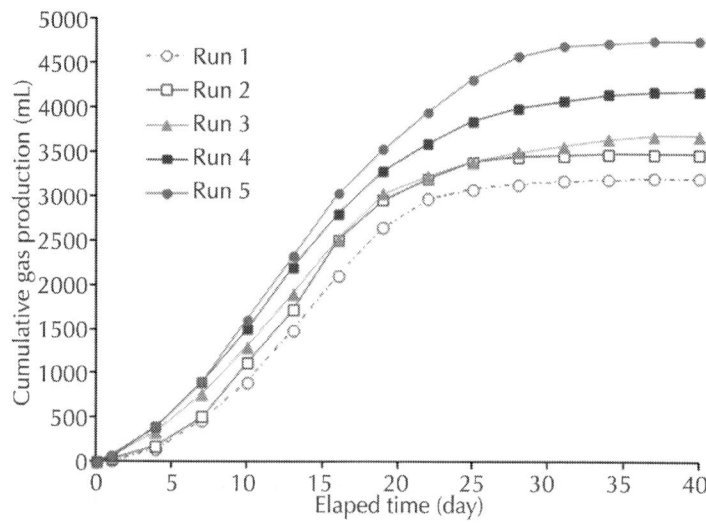

Figure 2: Comparison of cumulative gas yield from five types of samples.

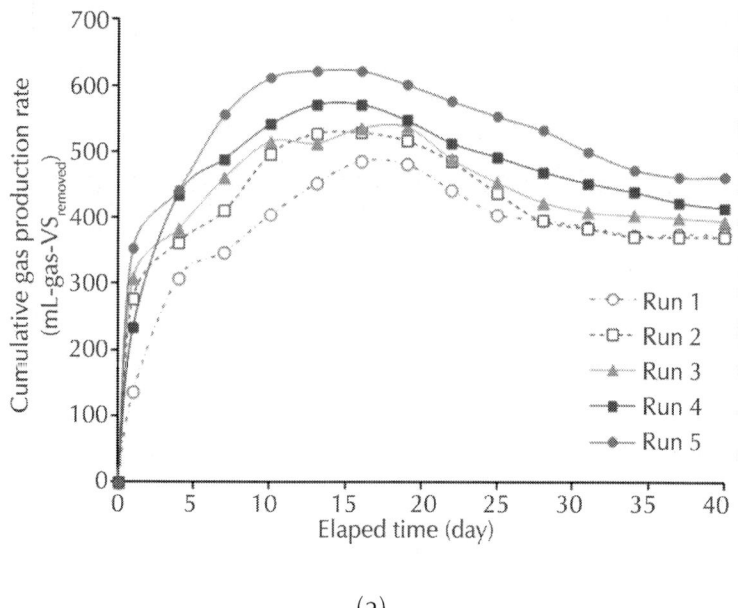

(a)

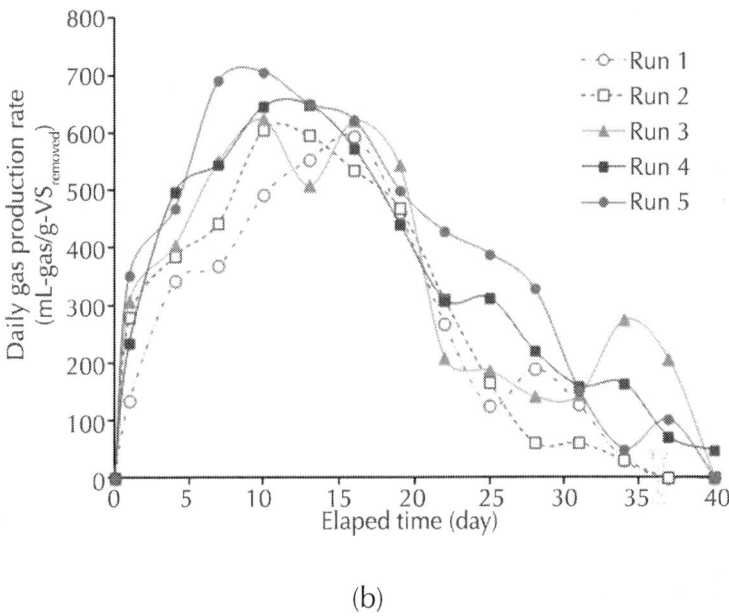

(b)

Figure 3: Distribution of gas production rate from five types of samples (3 days period) (a) Cumulative gas production rate; (b) Daily gas production rate.

Modified Gompertz Model (MGM) and Exponential Model (EM) were used to evaluate the cumulative methane yield on a basis of the BMP assay operated for 40 days under the experimental conditions for five sample groups. By comparison of the ultimate methane yields (cumulative methane yields at the final time, 40 days) evaluated by two models, MGM (168.39 - 314.08 mLCH$_4$/g-VS$_{added}$) determined relatively lower ranges than EM (167.62 - 312.47 mL-CH$_4$/g-VS$_{added}$).

In Table 6, for the mesophilic-dry digestion experiment, the ultimate methane yield was observed in the range of 163.02 mL - 313.27 mL-CH$_4$/g-VS$_{added}$ which was applied in two models as M_o and B_o. The methane yield was also increased when the chaff mixing ratio was high as shown in the case of total gas production rate. Also, Run 1 without the trace elements indicated the lower cumulative methane yield (77.15 mL-CH$_4$/g-VS$_{added}$) than that for Run 2 with the trace elements.

Table 6: Summary of kinetic parameters predicted by two models

Model	Parameters	Run 1	Run 2	Run 3	Run 4	Run 5
Modified Gompertz Model Ultimate methane yield, M_o (mL-CH_4/g-VS_{added}) Maximum methane production rate, R_m (mL-CH_4/g-VS_{added}·day) Lag phase, λ (day) Determination coefficient, R^2 (-)	Cumulative methane yield, M (mL-CH_4/g-VS_{added})	168.49	245.33	256.65	291.22	314.08
		168.02	245.17	256.03	291.57	313.27
		15.98	16.97	16.90	17.58	19.77
		3.0	3.0	3.0	3.0	3.0
		0.94	0.98	0.99	0.99	0.98
Exponential Model Ultimate methane yield, B_o (mL-CH_4/g-VS_{added}) Reaction rate constant, k (/day) Determination coefficient, R^2 (-)	Cumulative methane yield, B (mL-CH_4/g-VS_{added})	167.62	245.04	254.79	291.14	312.47
		168.02	245.17	256.03	291.57	313.27
		0.15	0.19	0.13	0.16	0.15
		0.89	0.84	0.92	0.85	0.87

Note) *M_o or B_o is the ultimate methane yield observed during the experiment.

The results of BMP assay simulated by two types of models showed a difference and the modified Gompertz Model was better fit compared with the Exponential Model as shown in Figure 4.

In the experiment, the ultimate methane yield per the inserted VS was distributed in the range of 163.02 mL - 313.27 mL-CH_4/g-VS_{added}. In previous studies, mixture digestion of food waste and livestock manure was tested in the thermophilic-wet digestion process and the ultimate methane yield per the inserted VS was ranged from 313.35 to 377.43 mL-CH4/g-VS_{added} [16], which was relatively higher than that obtained in this study, and similar to that of the single livestock manure tested at the thermophilic digestion and its result presented as 241 mL/g-VS$_{added}$ [17]. In addition, typical experimental values obtained from foreign countries were 250 L/g-VS for the cow manure including the straw and 279 L/g-VS [18] for the horse-manure with the straw estimated by the thermophilic-wet anaerobic digestion process, and 318 mL/g-VS$_{added}$ [19] for sewage sludge in similar. However, comparative evaluation represented little because BMP assay were varied greatly due to the operation conditions such as nutriment component and addition amount, heat, agitation, and so on [20].

According to the evaluation result of the ultimate methane yield, the mixture of the livestock manure and the chaff in the mesophilic anaerobic digestion process has an advantage when comparing with the only livestock manure containing high moisture content at the thermophilic digestion process in common. Also, it is verified that the trace elements are required to add when the chaff to mix with the livestock manure is not sufficient. Therefore, further study is required more detailed research.

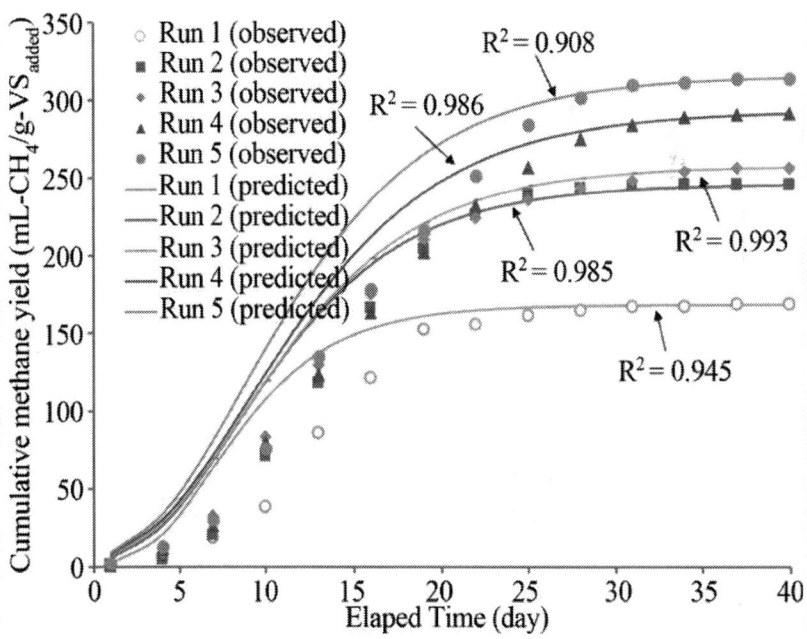

(a)

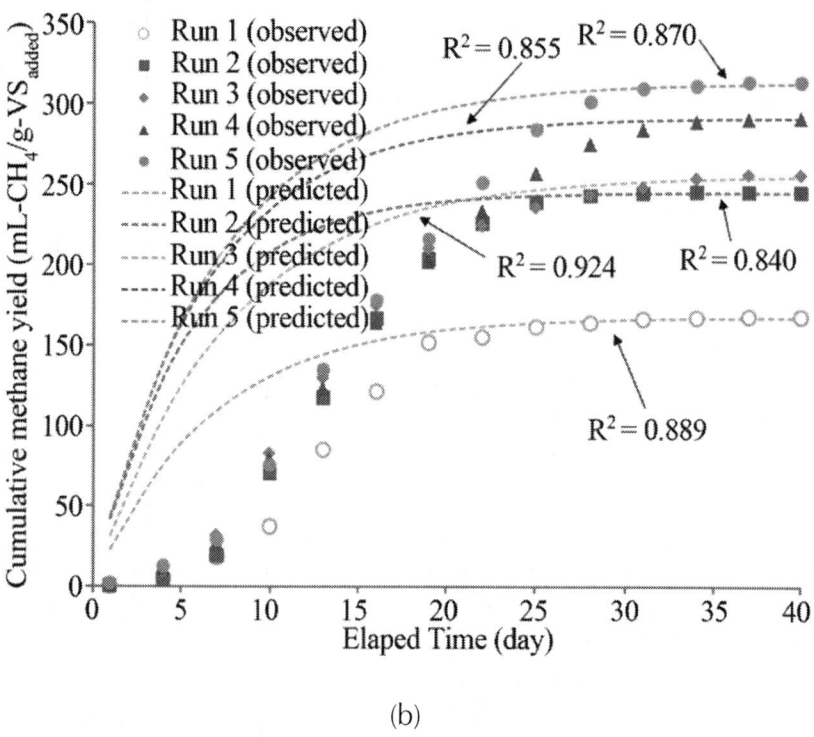

(b)

Figure 4: Biochemical methane potential assay simulated by two models. a) Modified Gompertz Model; b) Exponential Model.

Mixture Ratio of Manure and Chaff

On the domestic side, food waste as a mixture substrate of livestock manure has been mainly used to solve a pending issue and to improve methane production efficiency. Domestic studies related to optimal mixing ratio of the livestock manure and the food waste have been conducted but the study using herbal plants such as chaff and straw is rare. Because rural areas mainly produce the livestock manure and urban areas discharge the food waste mostly, facilities and energy for transportation are required if the food waste is considered as the substrate mixing with the animal manure. The herbal plants like the chaff are easily obtained in the rural area and mostly used to mix with the livestock manure as shown in developed countries having many cases for biogas plants [21].

With that background, this study verified that the methane production rate was varied due to the amount of the chaff mixed with the livestock manure. As the amount of the chaff was enlarged to expand biogas production and to increase methane production efficiency, it was naturally converted to the dry anaerobic digestion process caused by the lower moisture content and it had little regard for wastewater emission. Coverse et al. studied on the methane production using the high rate of organic loading as inserting carbohydrate mixture compounds [22]. According to their study, the maximum methane production rate was high at the thermophilic digestion but overall methane production yield was larger at the mesophilic digestion. A previous study applied the straw as the mixture substance had reported the most methane production was yielded from the mixture sample of 3% straw and the livestock manure containing 5% solid [23].

To verify the optimal mixing ratio of the piggery manure (SM) as the main compound and the chaff (SB) as the additional mixing compound in this study, the ultimate methane production yield and the maximum production rate due to SM/SB ratio are presented in Figure 5. As shown in Figure 5(a), the ultimate methane production yield and the maximum methane production rate were increased less than 3.5 of SM/SB ratio. The SM/SB ratio is remarkably varied due to the livestock manure composition changed by the stock farmers and the discharging season (or time) and its role may be slightly changed due to the type of the chaff and the condition of the chaff. Therefore, it is necessary to develop a standard model using the more accurate SM/SB ratio examined by several parameters such as solid content, livestock manure types, chaff types, dry conditions of the chaff including moisture content and woody presence, and so on.

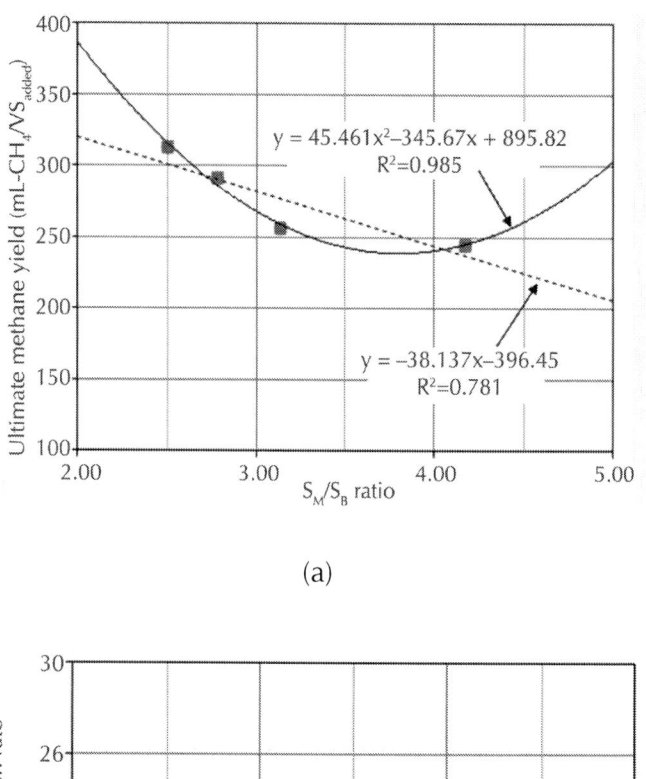

(a)

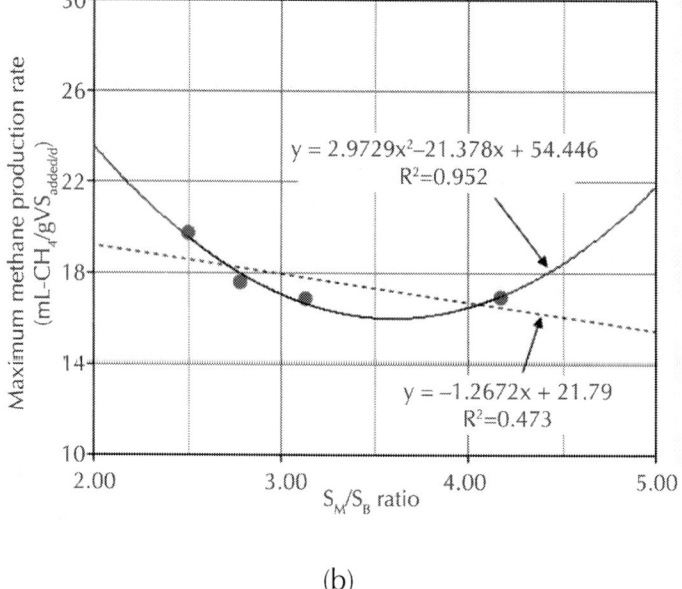

(b)

Figure 5: Change of methane yield coefficients in terms of M/H ratio a) Ultimate methane yield; b) Maximum methane production rate.

CONCLUSIONS

Considering domestic waste and natural water management, piggery manure is mostly required to be treated among non-point sources of the stock raising system and can also be a main compound of methane production. With that background, this study performed a series of experiments on mesophilic-dry anaerobic digestion of the main compound, piggery manure, mixed with chaff, a mixture compound or substrate, for supplying carbon and controlling moisture. Also, this study conducted comparative evaluation of methane production potential produced from the anaerobic digestion process due to a mixing ratio of the piggery manure and the chaff, and obtained following conclusions after the careful consideration and emphasis on the reasonable mixing ratio:

- The larger amounts of the chaff resulted in the larger methane production yields as well as the larger gas production yields per volatile solid (VS) inserted. The methane ratio in the produced biogas was also increased as the amount of the chaff was increased.
- When using the same mixture ratio of the chaff, the gas production from the sample without trace elements was less than that from the sample with trace elements. Therefore, one can say that the trace elements are important when producing the methane from the mixture compounds of the piggery manure and the chaff.
- For the mesophilic-dry digestion, the cumulative methane production yield per the inserted VS was ranged in 163.02 mL - 313.27 mL-CH4/g-VS_{added}. Similar to the measurement result of total gas production yield, the methane production yield was increased with the higher mixture ratio of the chaff.
- For the comparative evaluation of the BMP assay, the ultimate methane yield using Modified Gompertz Model was calculated as 168.39 - 314.08 mL-CH_4/gVS_{added}. It was slightly lower than that in the thermophilic-wet digestion process for the mixture of the domestic food waste and livestock manure and similar to that in the thermophilic-wet digestion process of the sewage activated sludge.
- When the optimal mixture ratio (SM/SB) between the piggery manure (SM) and the chaff (SB) was less than 3.5, the ultimate

methane production yield (M_o) and the maximum production rate (R_m) were decreased.

In conclusion, the mixture of the chaff to the piggery manure controls the moisture content and improves the methane production rate. Also, this study verifies the feasibility of the mesophilic-dry anaerobic digestion process, which is less difficult in operating and maintaining.

ACKNOWLEDGEMENTS

This study was supported by a grant (CUHBRI-2012-0 2-007) of the CNUH-BRI as the project, Development of Future Oriented Healthcare Model and Fundamental Technology for Agro-Medical System, and partly supported by a grant (2013 research project) of Jeonbuk Green Environment Center. Also, we appreciate to Dr. Kyung-Yub Hwang (Research Fellow Emeritus) in Korea Institute of Science and Technology who gives us professional advice to complete this paper and experiments.

REFERENCES

1. P. Börjesson and B. Mattiasson, "Biogas as a ResourceEfficient Vehicle Fuel," Trends Biotechnology, Vol. 26, No. 1, 2008, pp. 7-13.http://dx.doi.org/10.1016/j.tibtech.2007.09.007
2. H. J. Park, M. K. Song and C. K. Na, "Pretreatment Efficiency of Piggery Wastewater Using Coagulation-MAP Sedimentation," Journal of Korea Society of Waste Management, Vol. 27, 2010, pp. 457-466.
3. Y. M. Yoon, Y. J. Kim and C. H. Kim, "The Evaluation of Economic Efficiency to Composting and Liquefying Process of Biomass Discharged in Pig Breeding," Agriculture Economics, Vol. 31, 2009, pp. 39-62.
4. D. Bolzonella, L. Innocenti, P. Pavan, P. Traverso and F. Cecchi, "Semi-Dry Thermophilic Anaerobic Digestion of the Organic Fraction of Municipal Solid Waste: Focusing on the Start-Up Phase," Bioresource Technology, Vol. 86, No. 2, 2003, pp. 123-129.http://dx.doi.org/10.1016/S0960-8524(02)00161-X

5. P. Pavan, P. Battistoni and J. Mata-Alvarez, "Performance of Thermophilic Semi-Dry Anaerobic Digestion Process Changing the Feed Biodegradability," Water Science and Technology, Vol. 41, 2000, pp. 75-81.
6. N. Forster-Carneiro, M. Perez and L. I. Romero, "Anaerobic Digestion of Municipal Solid Wastes: Dry Thermophilic Performance," Bioresource Technology, Vol. 99, No. 17, 2008, pp. 8180-8184. http://dx.doi.org/10.1016/j.biortech.2008.03.021
7. B. Montero, J. L. Garcia-Morales, D. Sales and R. Solera, "Analysis of Methanogenic Activity in a Thermophilicdry Anaerobic Reactor: Use of Fluorescent in Situ Hybridization," Waste Management, Vol. 29, No. 3, 2009, pp. 1144-1151. http://dx.doi.org/10.1016/j.wasman.2008.08.010
8. S. E. Oh, M. K. Lee and D. H. Kim, "Continuous Mesophilic-Dry Anaerobic Digestion of organic Solid Waste," Journal of Korean Society of Environmental Engineers, Vol. 31, 2009, pp. 341-345.
9. J. J. Lay, Y. Y. Li and T. Noike, "Development of Bacterial Population and Methanogenic Activity in a Laboratory-Scale Landfill Bioreactor," Water Research, Vol. 32, No. 12, 1998, pp. 3673-3679. http://dx.doi.org/10.1016/S0043-1354(98)00137-7
10. W. F. Owen, D. C. Stuckey, J. B. Healy, L. Y. Young and P. L. McCarty, "Bioassay for Monitoring Biochemical Methane Potential and Anaerobic Toxicity," Water Research, Vol. 13, No. 6, 1979, pp. 485-492. http://dx.doi.org/10.1016/0043-1354(79)90043-5
11. R. S. Daniel and J. M. Tiedje, "General Method for Determining Anaerobic Biodegradation Potential," Applied and Environmental Microbiology, Vol. 47, 1984, pp. 850-857.
12. I. Angelidaki, M. Alves, D. Bolzonella, L. Borzacconi, J. L. Campos, A. J. Guwy, S. Kaalyuzhnyi, P. Jenicek and J. B. van Lier, "Defining the Biomethane Potential (BMP) of Solid Organic Wastes and Energy Crops: A Proposed Protocol for Batch Assays," Water Science and Technology, Vol. 59, No. 5, 2009, pp. 927-934. http://dx.doi.org/10.2166/wst.2009.040
13. M. H. Zwietering, I. Jongenburger, F. M. Rombouts and K. van't Riet, "Modeling of the Bacterial Growth Curve," Applied and Environmental Microbiology, Vol. 56, No. 6, 1990, pp. 1875-1881.

14. T. L. Hansen, J. E. Schmidt, I. Angelidaki, E. Marca, J. Cour Jansen, H. Mosboek and T. H. Christensen, "Method for Determination of Methane Potentials of Solid Organic Waste," Waste Management, Vol. 24, No. 4, 2004, pp. 393-400. http://dx.doi.org/10.1016/j.wasman.2003.09.009
15. H. B. Møllera, S. G. Sommera and B. K. Ahringb, "Methane Productivity of Manure, Straw and Solid Fractions of Manure," Biomass and Bioenergy, Vol. 26, No. 5, 2004, pp. 485-495. http://dx.doi.org/10.1016/j.biombioe.2003.08.008
16. J. K. Park, S. R. Jeong, J. H. Kang, Y. M. Ahn, H. E. Jin and N. H. Lee, "A Study on Optimization Condition for Anaerobic Co-Digestion of Food Waste with Livestock Wastes," Journal of Korea Society of Waste Management, Vol. 29, 2012, pp. 356-364.
17. S. H. Kim, H. C. Kim, C. H. Kim and Y. M. Yoon, "The Measurement of Biochemical Methane Potential in the Several Organic Waste Resources," Korean Journal of Soil Science and Fertilizer, Vol. 43, 2010, pp. 356-362.
18. S. Aslanzadeh, M. J. Taherzadeh and I. S. Horvath, "Pretreatment of Straw Fraction of Manure for Improved Biogas Production," Bio-Resources, Vol. 6, 2011, pp. 5193-5205.
19. J. G. Lin, Y. S. Ma, A. C. Chao and C. L. Huang, "BMP Tests on Chemically Pretreated Sludge," Bioresources Technology, Vol. 68, No. 2, 1999, pp. 187-192. http://dx.doi.org/10.1016/S0960-8524(98)00126-6
20. P. Shanmugam and N. J. Horan, "Simple and Rapid Methods to Evaluate Methane Potential and Biomass Yield for a Range of Mixed Solid Wastes," Bioresource Technology, Vol. 100, No. 1, 2008, pp. 471-474. http://dx.doi.org/10.1016/j.biortech.2008.06.027
21. D. Jackowiak, D. Bassard, A. Pauss and T. Ribeiro, "Optimization of a Microwave Pretreatment of Wheat Straw for Methane Production," Bioresource Technology, Vol. 102, No. 12, 2011, pp. 6750-6756. http://dx.doi.org/10.1016/j.biortech.2011.03.107
22. J. C Converse, R. E. Graves and G. W. Evans, "Anaerobic Degradation of Dairy Manure under Mesophilic and Thernophilic Temperatures," Transactions of the ASAE, Vol. 20, 1977, pp. 336-340.

23. J. E. Robbins, M. T. Armold and S. L. Lacher, "Methane Production from Cattle Waste and Delignified Strawt," Infection and Immunity, Vol. 38, 1979, pp. 175-177.

Chapter 10

Simulation of CO_2 and H_2S Removal Using Methanol in Hollow Fiber Membrane Gas Absorber (HFMGA)

Majid Mahdavian[1], Hossein Atashi[1], Morteza Zivdar[1], and Mahmood Mousavi[2]

[1]Department of Chemical Engineering, University of Sistan and Baluchestan, Zahedan, Iran

[2]Department of Chemical Engineering, Ferdowsi University of Mashhad, Mashhad, Iran

ABSTRACT

Application of methanol solvent for physical absorption of CO_2 and H_2S from $CO_2/H_2S/CH_4$ mixture in gas-liquid hollow fiber membrane gas absorber (HFMGA) was investigated. A computational mass transfer

(CMT) model for simulation of HFMGA in the case of simultaneous separation of CO_2 and H_2S was developed. The membrane gas absorber model explicitly calculates for the rates of mass transfer through the membrane and components concentration profiles. Due to the lack of experimental data in the literature, the model was validated using available individual components' water absorption data. The numerical predictions were in good agreement with the experimental data. The effects of operating conditions such as liquid velocity, gas velocity, temperature and pressure were analyzed. It is shown that methanol solvent can successfully be used for CO_2 and H_2S removal in membrane gas absorber. Also it is found that the concentration distribution of CO_2 and H_2S in the gas phase along the fiber length obeys plug flow model whereas in the methanol absorbent deeply affected by the interface concentration, absorbent velocity and diffusivity. In addition, it is shown that application of membrane gas absorber using methanol absorbents for H_2S removal and at higher flow rate is more efficient. Moreover, at operating pressures above 10 atm even at low absorbent rate, H_2S concentration depletion is relatively complete while at 1 atm this value is about 30%. This means that removal efficiency decreases with an increase in temperature and it is more important especially for H_2S.

INTRODUCTION

Some industrial gas streams (such as natural gas processing, petroleum refineries, petrochemicals) frequently contain H_2S and CO_2 as impurities. All of these gases requires treatment before delivery to the pipeline. It is reported that CO_2 is representing about 80% of greenhouse gases and half of the CO_2 emissions are produced by industrial plants such as fossil-fuel-fired power plants, iron, steel and cement works [1]. Also carbon dioxide is a common contaminant of natural gas and must be removed to a level of <8% (usually <2%) to minimize corrosion of the pipeline. Hydrogen sulfide removal is also desirable to reduce corrosion. In many cases it is necessary from the health and safety standpoint [2].

The most well known technology for recovery/removal of CO_2 and H_2S is solvent absorption. This technology was established over 80 years ago in the chemical and oil industrials for the removal of acid gases from natural gas streams. For the removal of CO_2 and H_2S, traditionally absorption processes like packed and plate columns are utilized [3]. Because these generally require large space and high investment cost, the emphasis of designing most of these operations is towards maximizing the mass transfer rate by creating as much interfacial area as possible [4]. In addition, they also suffer from several limitations including flooding, loading entrainment, foaming, weeping, etc. In recent years, the demand for alternative technologies has increased and many researchers have looked for new technologies to enhance the efficiency of absorption processes. Membrane-based absorption technique has been introduced as an emerging technology for the recovery/removal of gases (like CO_2, H_2S, SO_2, NH_3, VOC, etc.) from various industrial process gas streams [5]. In addition to gas/liquid, this technology also has found applications in numerous liquid/liquid applications such as fermentation, pharmaceuticals, wastewater treatment, semiconductor manufacturing, carbonation of beverages, metal ion extraction, protein extraction, osmotic distillation and other operations [6].

Membrane gas absorbers are devices that achieve twophase mass transfer through diffusion without dispersing one phase within another. Such a device employs a porous membrane acts as a non-selective barrier between both phases where the gas and the absorbent solution flow on two sides of a membrane [5]. The membranes are usually microporous and can be both hydrophobic and hydrophilic. Hydrophobic microporous membranes like polypropylene (PP), polytetrafluoroethylene (PTFE) and polyvinylidene fluoride (PVDF) membranes have received increasing attention in recent years for using in membrane gas absorbers because of their good hydrophobicity [7]. These membrane absorber systems, generally in the form of hollow fibers with diameters of 0.5 mm - 1 mm in densely packed membrane modules, provide a high interfacial area (500 m^2/m^3 - 2000 m^2/m^3) significantly greater than most traditional absorbers (100 m^2/m^3 - 800 m^2/m^3) between two phases to achieve high overall rates of mass transfer. This significantly decreases the size required for the contactor [8]. Moreover, this kinds of contacting devices offers a number of important advantages over conventional dispersed phase contactor for gas sorption, such as large interfacial area between gas and liquid flow (up to two orders of magnitude more surface area per volume than conventional contactors), no flooding and foaming phenomena, independent control of gas and liquid flow rates, high efficiency, the possibility of combining absorption and desorption in one single compact module, energy intensive, and so on (as an example of review, see Gabelman et al. [5]).

Chemical absorbents like amines and amino acid salts are extensively used in the removal of impurities from gas mixtures. Physical absorbents have been of considerable interest in the development of gas treatment solvents, especially when the partial pressure of undesirable impurity is high. Some of the physical solvents used commercially are propylene carbonate (PC), n-formyl morpholine (NFM), dimethyl ethers of polyethylene glycol (DEPG), and n-methyl-pyrrolidone (NMP) (see more example in [9]). Physical solvents can be a possible alternative to chemical solvents in certain areas of applications, although they are less effective than chemical absorbents (i.e. the specific absorption rate into physical absorbents in comparison with chemical solvents is less). But they can be regenerated by just pressure reduction method without large amount of heat supply and thus excessive energy savings can be obtained [9]. An economical analysis must be done to select the best choice of solvent. In addition, they can be used as pre-treatment solvent in the development of hybrid systems. The most well-known physical absorbent is water. However, its economics are limited by the relatively low solubility which leads to larger amounts of circulation rate, i.e. the higher investment costs as well as the higher operating costs [10]. However, there are good organic solvents which possess a much higher solvent capacity than water. Among the physical solvents, n-methyl-2-pyrrolidone (NMP), methanol and propylene carbonate (PC) are popular as gas treating solvents. Methanol has a high thermal and chemical stability, low vapor pressure, and is not corrosive. It is able to absorb acid gases, hydrocarbons, mercaptans and water. Moreover, it is produced in big quantity and readily available [10]. This properties make it highly effective for processing a wide range of compositions.

The applications of hollow fiber gas-liquid membrane gas absorber for acid gas removal specially carbon dioxide from gas mixtures have been studied by several researchers. In this case, a large number of experimental absorption studies and theoretical modeling analyses have been performed with physical or chemical absorbent liquids such as pure water, aqueous amine solution, aqueous sodium hydroxide solution, aqueous potassium carbonate solution, aqueous blended solvents, etc. [11- 14]. Some authors have explored possible simultaneous removal of H_2S and CO_2 in hollow fiber membrane gas absorber using MEA [1] and DEA [15]. However, of the authors considering chemical absorption, few have worked with physical solvents as would be the good choice in membrane gas absorber process. There have been few attempts to address possible physical absorption in hollow fiber membrane gas absorbers [16-18] that mostly describes the water performance and theoretical analysis of simultaneous removal of CO_2 and H_2S using methanol absorbent in HFMGA has not been discussed by researchers.

In the present work, after modification of 2D mathematical model, this new process has been applied for CO_2 and H_2S capture from carbon dioxide/hydrogen sulfide/ methane mixture (when the partial pressures of CO_2 and H_2S are 10% of total pressure) using methanol (as an example of physical absorbent) absorbent and its potential possibility for carbon dioxide and hydrogen sulfide removal has been evaluated. It should be mentioned that areas of possible HFMGA process for gas treatment using physical solvent with economic considerations will be reported in another work. This work was performed using CFD tool with respect to solubility behavior. CFD has been largely used as a powerful tool to model membrane separation processes. It is able to simulate the concentration, temperature and velocity fields as well as the transport parameters and operating efficiency.

MODEL DEVELOPMENT

In this paper, a steady-state two-dimensional mathematical model has been modified (e.g. [1,12]) to describe the physical absorption of carbon dioxide and hydrogen sulfide in the polymeric hollow fiber membrane gas absorber (using methanol absorbent as the absorption liquid). The model describes the mass transfer in the gas, membrane and liquid phases. Axial and radial diffusion inside the shell, through the membrane, and within the tube side of the membrane gas absorber have been considered in the model equations. It allows studying the effect of membrane wetting on the mass transfer through the membrane and also the effect of operating conditions (gas and liquid flow rates, temperature), solvent affinity (H) and flow pattern (counter current or co current arrangement) on the carbon dioxide and hydrogen sulfide removal efficiencies.

This model assumes that the fibers are distributed evenly through the shell space, which allows the results obtained with a single fiber to be generalized to the entire module. Model results are based on "non-wetted mode" in which the gas mixture filled the membrane pores.

The following assumptions are made to develop the governing mass transfer differential equations: 1) fully developed parabolic velocity profile in the hollow fiber under laminar flow conditions; 2) the mixture gases flow inside the shell are ideal gas; 3) Happel's free surface model [19] is used to characterize the velocity profile at the shell side; 4) the physical properties of the fluid are constant; 5) the Henry's law is applicable for gas-liquid interface; 6) no absorption of bulk and inert gases; 7) pitch and placing of the fibers are uniform; 8) no pore blockage.

Transport Model for the Hollow Fiber Membrane Gas Absorber

In order to describe the mass transfer and develop the equations of mathematical model in the hollow fiber membrane gas absorber, a material balance has been applied for a segment of a hollow fiber, as shown in Figure 1 in the shell, membrane and tube sides. Also, the computational domain used for the numerical simulation is shown in Figure 1. This model is based on the idea that two concentric cylinders are used as the model for fluid flowing out of the fibers and so only portion of fluid surrounding the fiber is considered and may be approximated as circular cross section [19] The fluid flow is described using the fully developed laminar flow model in the tube side, whereas the fluid flow in the shell side is characterized by the Happel's free surface model.

The position $r = 0$ is the center of the fiber and r_1, r_2 and r_3 are the inner, outer and Happel's free model radii of the fiber, respectively (Figure 1). The radius of Happel's free surface model is calculated to be $r_3 = 720$ µm. Dimensions of the hollow-fiber membrane gas absorber are listed inTable 1. The gas mixture consists of carbon dioxide, hydrogen sulfide and methane is fed to the shell side at z =

L, while the liquid (methanol) is passed through the tube side at z = 0. Carbon dioxide and hydrogen sulfide are removed from the mixture by diffusing through the membrane due to a concentration gradient and then absorbing with the solvent.

Equations Describing the Shell Side

Convective-diffusion equation for the component i using Fick's law of diffusion, when chemical reaction is taking place, can be written as:

$$\partial C_{i,shell}/\partial t + \nabla \cdot \left(-D_{i,shell}\nabla C_{i,shell} - C_{i,shell}V_{i,shell}\right) = R_{i,shell} \quad (1)$$

Where C_i, $D_{i,shell}$, R_i and V_{shell} denote the local concentration of the component i, the diffusivity of the component i, reaction rate of the component i and axial velocity in shell side, respectively. According to the Happel's free surface model [19], the velocity profile in the shell side may be obtained. For this purpose, a momentum balance over a thin cylindrical shell is integrated twice to obtaining the following equation for the shell side velocity distribution, which have been applied by several authors (e.g. [20,21]):

$$V_{shell} = 2V_{ave-shell}\left(1-\left(\frac{r_2}{r_3}\right)^2\right)$$

$$\cdot \left(\frac{(r/r_3)^2 - (r_2/r_3)^2 + 2\ln(r_2/r)}{3+(r_2/r_3)^4 - 4(r_2/r_3)^2 + 4\ln(r_2/r_3)}\right) \quad (2)$$

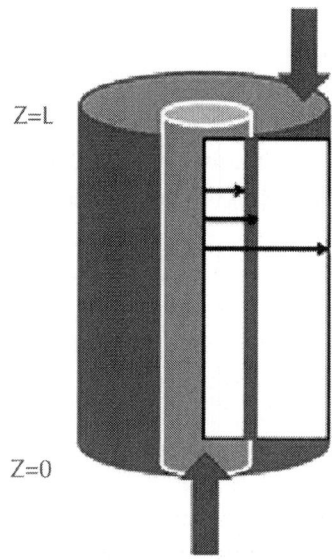

Figure 1: The schematic diagram of a hollow fiber membrane and computational domain.

Table 1: Specifications of the membrane gas absorber

Parameter	Value
Module I.D. (m)	4.35×10^{-3}
Module O.D. (m)	6.35×10^{-3}
Fiber O.D. (m)	9.07×10^{-4}
Fiber I.D. (m)	6.07×10^{-4}
Module length (m)	0.3
Fiber length (m)	0.2725
No. of fibers	9

where V_{shell} is the velocity component inside the shell, r_3 $V_{ave\text{-}shell}$ is the shell average velocity in the axial direction, r_2(m) is the outer fiber radius and (m) is Happel's free surface model radius defined as:

$$r_3 = r_2\sqrt{1/(1-\phi)} \tag{3}$$

Packing density (ϕ) can be defined as the ratio of total surface area of membrane to the cross-sectional area of the module and ϕ is calculated as:

$$\phi = n\left(\frac{r_2}{r_4}\right)^2 \tag{4}$$

where n is the number of fibers and r_4 is the inner radius of the MGA. The partial differential equation of the steady state mass balance for cylindrical coordinates, where no reaction takes place in the shell side is obtained and it is given as follows:

$$D_{i-shell}\left(\frac{\partial^2 C_{i-shell}}{\partial r^2} + \frac{1}{r}\frac{\partial C_{i-shell}}{\partial r} + \frac{\partial^2 C_{i-shell}}{\partial z^2}\right) = V_{shell}\frac{\partial C_{i-shell}}{\partial z} \tag{5}$$

The boundary conditions are the following (i = CO_2, H_2S)

At $z = L$: $C_{i-shell} = C_{0,i-shell}$ (6)

At $r = r_2$: $C_{i-shell} = C_{i-mem}$ (7)

At $r = r_3$: $\partial C_{i-shell}/\partial r = 0$ (symmetry) (8)

Equations Describing the Membrane Side

Mass transfer takes place through the membrane pores without mixing between phases and the transfer equation inside the pores can be derived without considering convection. The membrane diffusivity of species within the pores should be employed instead of the ordinary diffusivity. This parameter can be defined as $D_{i-mem} = D_i\varepsilon/\tau$. The steady state material balance for the transport of diffusing components (i = CO_2, H_2S) inside the membrane can be considered for non-wetting condition, where pores filled by the gas phase. For the non-wetting case, no chemical reactions will be considered in the membrane.

$$D_{i-mem}\left(\frac{\partial^2 C_{i-mem}}{\partial r^2} + \frac{1}{r}\frac{\partial C_{i-mem}}{\partial r} + \frac{\partial^2 C_{i-mem}}{\partial z^2}\right) = 0 \quad (9)$$

$$\text{At } r = r_1 : C_{i-mem} = C_{i-tube}/m_i \quad (10)$$

$$\text{At } r = r_2 : C_{i-mem} = C_{i-shell} \quad (11)$$

In the partial wetted mode, additional equations are required to describe diffusion-reaction inside the wetted parts of the pores. In this case the pores are both gas and liquid-filled (wetted and non-wetted parts) and the transport of the species i generally depends on its diffusion coefficient into the liquid [22].

Non-Wetted Part of the Membrane

$$D_{i-drymem}\left(\frac{\partial^2 C_{i-drymem}}{\partial r^2} + \frac{1}{r}\frac{\partial C_{i-drymem}}{\partial r} + \frac{\partial^2 C_{i-drymem}}{\partial z^2}\right) = 0 \quad (12)$$

Boundary conditions are:

$$\text{At } r = r_2 : C_{i-drymem} = C_{i-shell} \quad (13)$$

$$\text{At } r = r_w : C_{i-drymem} = C_{i-wetmem}/m_i \quad (14)$$

Wetted Part of the Membrane

$$D_{i-wetmem}\left(\frac{\partial^2 C_{i-wetmem}}{\partial r^2} + \frac{1}{r}\frac{\partial C_{i-wetmem}}{\partial r} + \frac{\partial^2 C_{i-wetmem}}{\partial z^2}\right) + \varepsilon R_i = 0 \quad (15)$$

Boundary conditions are:

$$\text{At } r = r_w : C_{i-wetmem} = m_i C_{i-drymem} \quad (16)$$

At $r = r_1 : C_{i-\text{wetmem}} = C_{i-\text{tube}}, C_{\text{solvent-tube}} = C_{\text{solvent-mem}}$ (17)

where m is dimensionless distribution coefficient.

Equations Describing the Tube Side

The partial differential equation of the steady state mass balance for each species during simultaneous mass transfer in a non-reactive absorption system is obtained and can be expressed as:

$$D_{i-\text{tube}} \left(\frac{\partial^2 C_{i-\text{tube}}}{\partial r^2} + \frac{1}{r} \frac{\partial C_{i-\text{tube}}}{\partial r} + \frac{\partial^2 C_{i-\text{tube}}}{\partial z^2} \right) = V_{\text{tube}} \frac{\partial C_{i-\text{tube}}}{\partial z} \quad (18)$$

The left-hand side of the above equation represents the diffusion and reaction terms, whereas the right-hand side is the convection term. Considering laminar flow, Navier-Stokes equations and the equation of continuity can be solved for fluid flow in a cylindrical pipe, therefore velocity distribution in the tube in the z direction can then be obtained as [22]:

$$V_{\text{tube}} = 2V_{\text{ave-tube}} \left(1 - \left(\frac{r}{r_1} \right)^2 \right) \quad (19)$$

Where V, r and r_1 are the average velocity in the lumen, the radial distance and the radius of the lumen, respectively. The following boundary conditions are considered:

At $z = 0 : C_{\text{solvent-shell}} = C_{0,\text{solvent-tube}}$ (20)

At $r = 0 : \partial C_{i-\text{tube}} / \partial r = 0 \text{ (symmetry)}$ (21)

At $r = r_1 : C_{i-\text{tube}} = m_i C_{i-\text{mem}}$ (22)

Physical Properties and Numerical Solution

Simulation of membrane gas absorber requires data on physicochemical properties used as input parameters in the model such as solubility and diffusivity of the relevant components in each phase. The distribution coefficient of CO_2 was taken from Versteeg et al. [23] and distribution coefficient of H_2S was taken from Carroll et al. [24] as a function of temperature in water. Henry's constant of CO_2 and H_2S for methanol as a function of temperature was reported by Lunsford et al. [25]. Liquidphase diffusivities of CO_2 [23] and H_2S [26] in water were estimated by the equations proposed by Versteeg and Cussler respectively and their value in methanol were estimated using the correlation given by Diaz et al. [27]. Gas-phase diffusivities of CO_2 and H_2S were estimated using the correlation given by Diaz et al. [27] and Cussler et al. [26]. Gas-filled membrane phase diffusivities were corrected for membrane porosity and tortuosity. The values for other data were obtained from [28,29].

In order to solve the coupled partial differential equations for the tube, membrane and shell sides with the appropriate boundary conditions and physical and chemical properties for CO_2 and H_2S, FEMLAB software has been used.

RESULTS AND DISCUSSION

Model Validation

The model of simultaneous absorption of CO_2 and H_2S using methanol in hollow fiber membrane gas absorber was validated using available individual components' absorption data of Al-Marzouqi et al. [20] and Faiz et al. [30] for physical absorption of CO_2 and H_2S in water, respectively, since we didn't find any reported experimenttal data for the type of the present work in the literature.

Comparison of the experimental data and the simulated results are shown in Figures 2 and 3. It should be noted that the type of hollow fiber MGA modules and the operating conditions applied for obtaining the mentioned experimental data differ significantly with each other and we applied the exact conditions for each case. Results of both validations for physical absorption of CO_2 and H_2S are shown in Figures 2 and 3, respectively. Figure 2 shows the calculation of the model with the experimental results of percent CO_2 removal as a function of water flow rate at the gas flow rate of 200 ml/min and Figure 3 shows the percent H_2S removal and the outlet H_2S concentration as a function of inlet concentration of H_2S in the gas phase at gas velocity of 5.1 m/s and liquid velocity of 0.092 m/s. Clearly the model predictions are in good agreement with the two set of CO_2 and H_2S absorption data and it shows that the numerical model accurately predicts the experimental data for both gases.

Concentration Distribution of CO_2 and H_2S

A steady state component concentration distribution is established inside the shell, membrane, and within the tube side of the hollow fiber membrane gas absorber that affects mass transfer coefficient, removal efficiency and mass transport. Numerically calculated carbon dioxide and hydrogen sulfide concentration distribution in each of the three phases are depicted in Figures 4 and 5, respectively. In order to compare the difference between concentration profiles of CO_2 and H_2S effectively, equal input concentration in gas phase is applied. The solubility of CO_2 and H_2S in methanol is linearly proportional to their partial pressure in the gas mixture and, hence, it can be modeled according to Henry's law. As expected, it can be seen that the concentration near the membrane-liquid wall signifycantly affected by the interface concentration, whereas the CO_2 and H_2S concentrations on the shell side slightly decrease in the radial direction. The concentration profile is discontinuous at the gas filled membrane-liquid interface based on the equilibrium relationship.

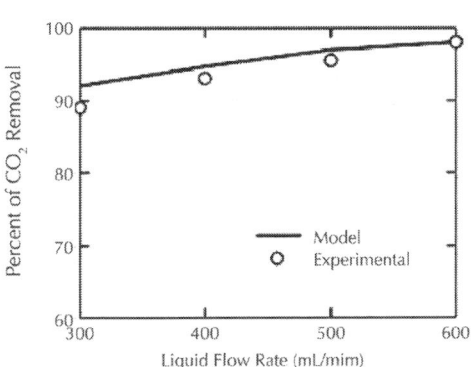

Figure 2: Comparison between experimental [20] and simulated CO_2 removal efficiency.

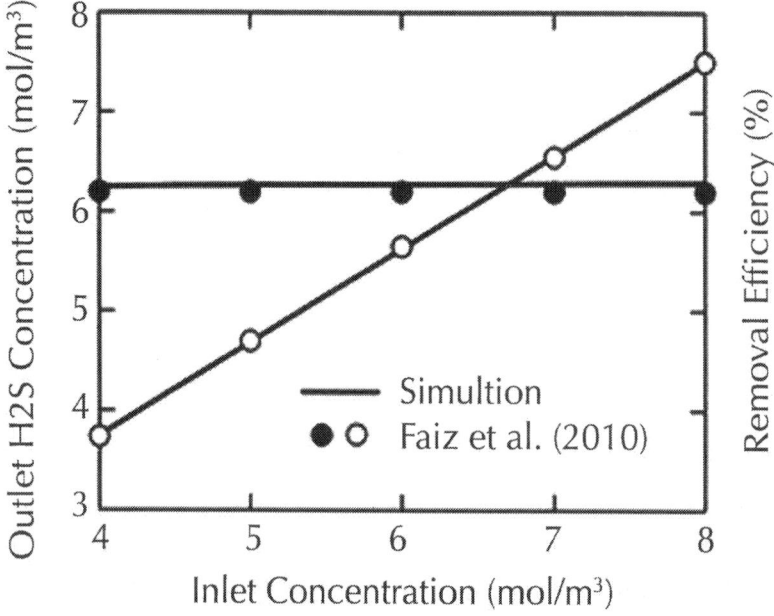

Figure 3: Comparison between model data [30] and simulated H_2S gas outlet concentration and removal efficiency.

Due to the dimensions of the hollow fiber, the computational domain is the area of membrane length multiplied by Happel's free surface model width. It is important to note that since the fiber is 900 times longer than its radial dimension (in this case, 0.3 mm in radius and 27 cm in length), a scaling factor of 90 has been applied in the z direction in order to reduce computational cost.

It is worth mentioning that the sensitivity grid-dependence analysis of the method of solution to the mesh size was performed in order to ensure that the numerical solution is not affected by the specification of the mesh size.

Simulation of CO_2 and H_2S Removal Using Methanol in Hollow ...

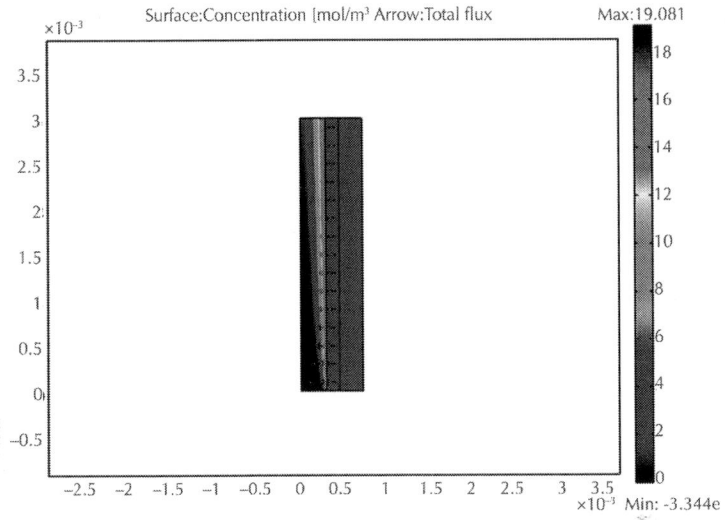

Figure 4: CO_2 concentration distribution in computational do main for V_L = 0.1 m/s, V_G = 3 m/s and T = 298 K.

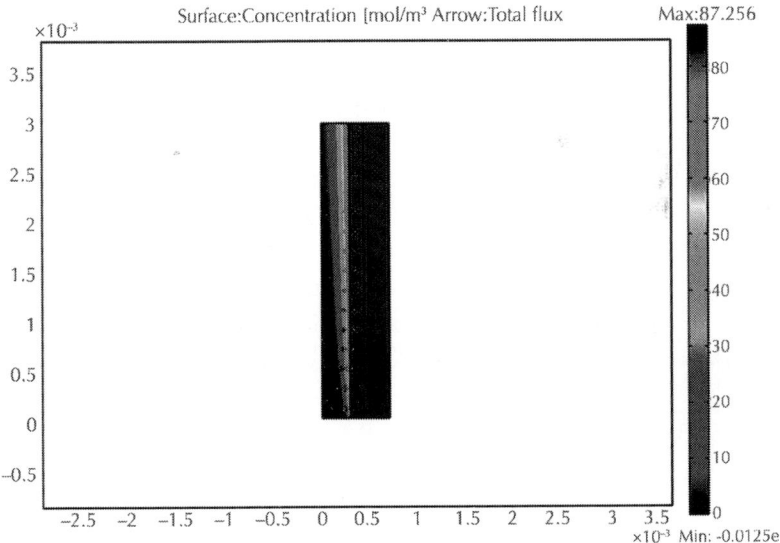

Figure 5: H_2S concentration distribution in computational domain for V_L = 0.1 m/s, V_G = 3 m/s and T = 298 K.

The importance of a fine mesh adjacent to the membrane wall is obvious from the components concentration distribution shown in Figures 4 and 5.

Figures 6 and 7 show the numerically calculated dimensionless radial concentration profile of CO_2 and H_2S as a function of dimensionless length at different cross sections along fiber length, i.e. z/L = 0.1, 0.5, 0.9 (Figure 6) and at two different absorbent velocities (Figure 7). With respect to these figures, there is a concentration drop close to the absorbent-membrane interface at the membrane wall in the methanol absorbent phase for both gases. Concentration depletion for CO_2 and H_2S in liquid phase has the same trend but there is a sharper reduction in H_2S concentration in comparison with CO_2 concentration which is attributed to higher solubility of H_2S in the methanol absorbent.

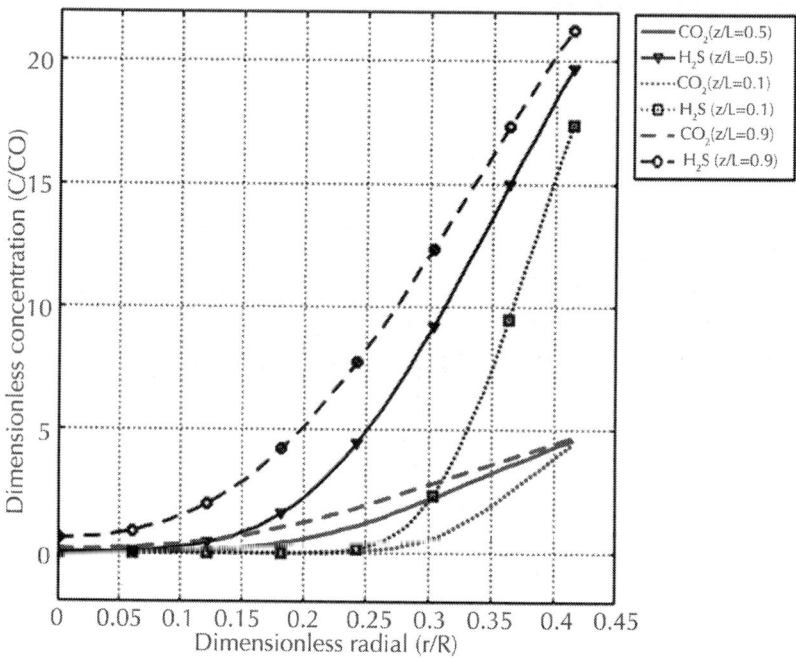

Figure 6: CO_2 and H_2S radial tube side concentration profiles along fiber length for z/L = 0.1, 0.5, 0.9, V_L = 0.1 m/s, V_G = 3 m/s and T = 298 K.

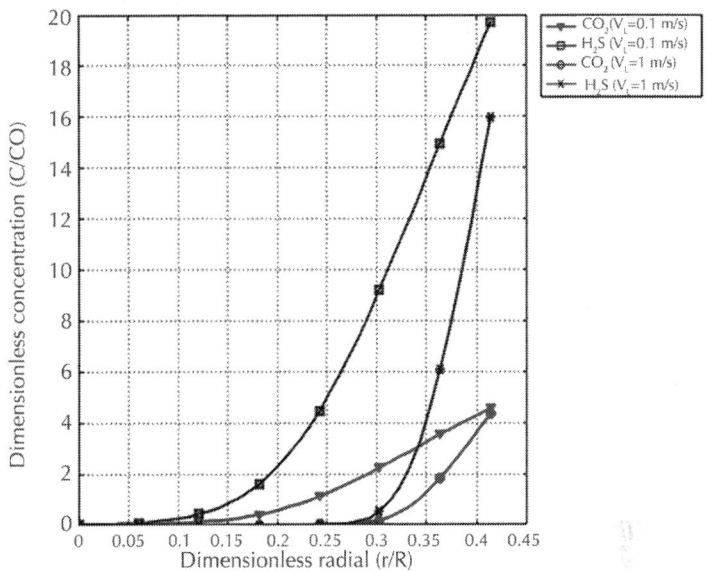

Figure 7: Effect of absorbent velocity on CO_2 and H_2S radial tube side concentration profiles for V_G = 3 m/s, z/L = 0.5 and T = 298 K.

The results indicate that penetration depth increases with distance from methanol absorbent entrance (z/L = 0) and, therefore, components diffuse into liquid phase from membrane interface. Note that since liquid phase is very thin, it acts as film layer. With respect to the components diffusivities, liquid velocity and dimension of fiber, it is seen that the contact time is not enough that diffusion entirely affects the liquid phase and absorbed species do not distribute rapidly in radial direction before absorbent leaves the fiber (dimensionless Gz number conception). However, at higher inlet liquid velocity, the depletion of component concentration is faster. The reason is that the axial convective flow decreases with radial diffusion.

Figure 8 shows the axial CO_2 and H_2S concentration profiles in absorbent and gas phase. For absorbent phase, tube center line (r = 0) and for gas phase Happel's radius (r = r_3) is selected. It can be seen that in the case of methanol absorbent, the CO_2 concentration depletion in the gas and amount of absorbed are low in comparison with the H_2S. It obviously indicates the higher capacity of methanol in absorption of H_2S. Based on the bulk concentration of CO_2 and

H_2S, removal efficiencies are about 12.9% and 29.3% throughout the fiber, respecttively. Moreover, in 50% and 40% of the fiber length, the concentrations are still zero for CO_2 and H_2S, respectively.

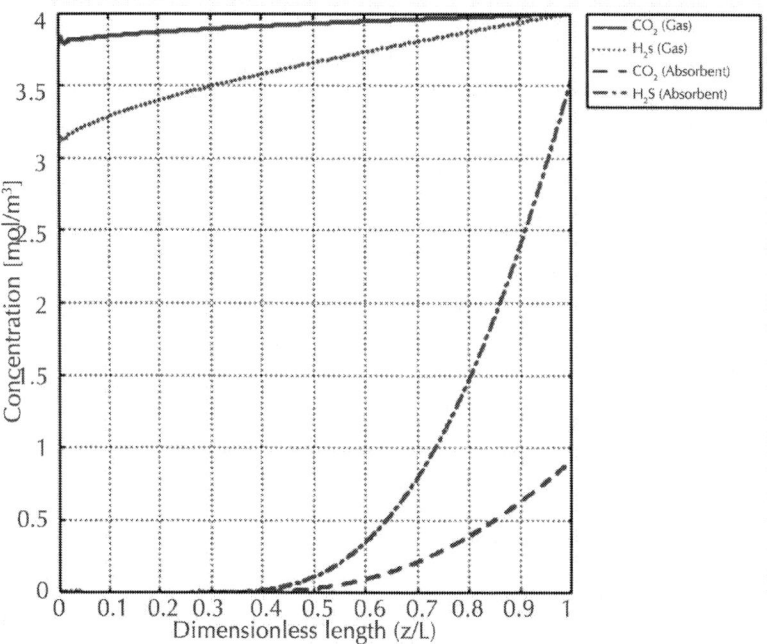

Figure 8: CO_2 and H_2S concentration profile in the axial direction in both shell and tube sides for V_L = 0.1 m/s, V_G = 3 m/s, C_{0,CO_2} = C_{0,H_2S} = 4 mol/m³ and T = 298 K.

Effect of Absorbent and Gas Velocity on CO_2 and H_2S Removal Efficiencies

Figures 9 and 10 indicate the effect of absorbent and gas velocity on the removal efficiencies of CO_2 and H_2S using methanol absorbent in comparison with water absorbent. Wide range of velocities was selected for both absorbent and gas in order to provide a chance to gain a real insight into this effect. Considering these figures, CO_2 and H_2S removal efficiencies at given conditions increase with the increase in absorbent velocity. This effect is due to the increasing in driving force with entering fresh absorbent. Therefore, CO_2 and H_2S concentrations in gas phase reduce and removal efficiencies improved because of higher absorption rate. This effect is reported by several authors [1,30,31] for water absorbent in the case of physical absorbent in hollow fiber membrane gas absorber devices.

The results show that with increasing the liquid velocity, the overall mass transfer coefficient increases. The reason is that in the case of physical absorption in membrane gas absorber, the controlling resistance for the mass transfer usually is liquid phase. The CO_2 and H_2S removal ability of methanol is illustrated in Figure 9 where the results are plotted for methanol in comparison with water absorbent. for example at the absorbent velocity of 3 m/s, the removal efficiencies using methanol absorbent are 34.7% and 84.3% and removal efficiencies using water absorbent are 13.8% and 21.6% for CO_2 and H_2S, respectively.

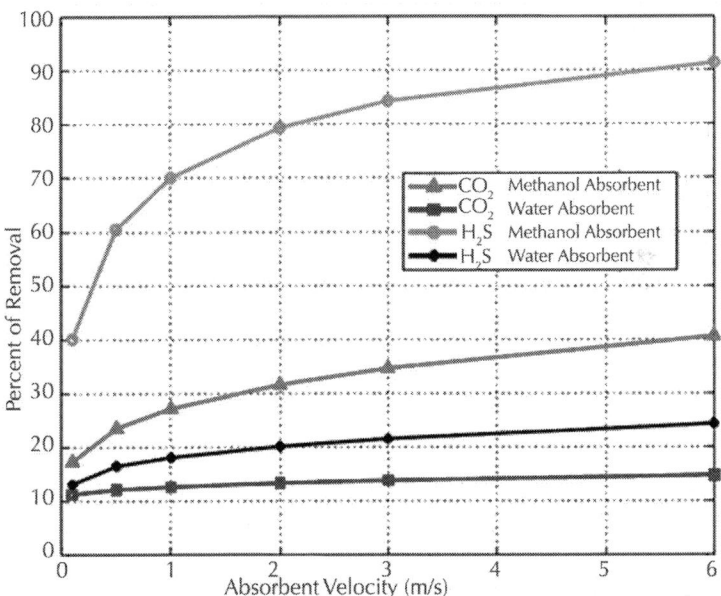

Figure 9: Effect of absorbent velocity on CO_2 and H_2S removal efficiencies for methanol and water absorbent at $V_G = 2$ m/s, $C_{0,CO_2} = C_{0,H_2S} = 4$ mol/m³ and T = 298 K.

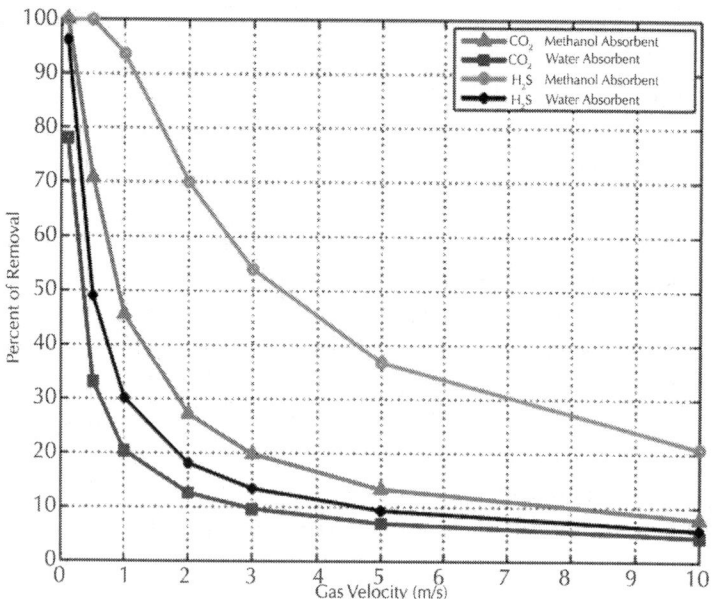

Figure 10: Effect of gas velocity on the CO_2 and H_2S removal efficiencies for methanol and water absorbent at $V_L = 1$ m/s, $C_{0,CO_2} = C_{0,H_2S} = 4$ mol/m^3 and $T = 298$ K.

It is important to note that in the case of methanol (or water), the CO_2 removal efficiency reaches a relatively constant value whereas, H_2S removal efficiency increases by increasing the liquid velocity which leads to higher relative absorption rates in comparison with CO_2. This is due to the fact that for higher absorbent velocities due to the lower contact time, the absorbent liquid cannot reach saturation and maybe leaves the module unsaturated. In spite of reducing contact time at higher velocities, methanol absorbent (or water) leaves the module saturated with respect to its low CO_2 potential absorption whereas the H_2S potential absorption is high enough resulting in an unsaturated absorbent at the module exit. Relative absorption rate of CO_2 using methanol absorbent is in the range of 1.5 to 2.7 and relative absorption rate of H_2S using methanol absorbent is in the range of 2.9 to 3.75 in comparison with the case of water absorbent in this operating. Therefore, application of membrane gas absorber using methanol absorbent for H_2S removal and at higher flow rate is more efficient. In addition, methanol in comparison with other commercially available physical solvents has a lower viscosity, which increases mass transfer rates and decreases membrane area requirements and pressure drop over the fiber length. Note that in simultaneous absorption of CO_2 and H_2S using methanol in MGA when selective absorption of CO_2 and H_2S is desired, selectivity remains relatively constant with increasing the methanol absorbent flow rate.

Figure 10 shows the effect of gas velocity on the CO_2 and H_2S removal efficiencies for methanol absorbent at a given conditions in comparison with water absorbent. It can be seen that CO_2 and H_2S removal efficiencies decrease considerably with the increase in gas velocity. This effect is due to the fact that by increasing the gas velocity (or flow rate) the amount of input impurity (CO_2 and H_2S) increases at constant absorption ability and on the other hand, gas-liquid contact time decreases. As a result of these two negative effects, CO_2 removal efficiency decreases in the membrane gas absorber.

Note that reduction in removal efficiencies are not the same for equal velocity step size in both gas and absorbent due to the different gas-liquid contact time. For example, contact time decreases 67% when velocity changes from 1 m/s to 3 m/s while it decreases 40% when velocity changes from 3 m/s to 5 m/s.

The Effect of Temperature and Pressure on CO_2 and H_2S Removal Efficiency

Figure 11 shows gas phase CO_2 and H_2S concentration profiles in the axial direction at three different temperatures, i.e. 288 K, 298 K and 308 K. It can be seen that the outlet CO_2 and H_2S concentrations increase and the trend of concentration variations for CO_2 and H_2S are the same but for H_2S is more important: the higher the temperature, the higher the average component concentration in the gas phase and outlet stream (lower removal efficiency). The reason is a result of two opposite effects that as the temperature increases, the solubility of CO_2 and H_2S decrease and liquid-phase diffusion coefficients increase. In addition, temperature effects on CO_2 and H_2S concentration distribution in the radial direction are more important near the membrane-liquid interface in the liquid phase.

Generally, physical solvents are used for undesirable component removal from high-pressure gas streams. Figure 12 shows the effect of pressure on CO_2 and H_2S removal efficiencies for methanol absorbent. In the case of application of membrane gas absorber at high pressures, methanol as a physical solvent is more efficient and it might be an alternative to chemical solvents. At the module exit, complete H_2S concentration depletion is relatively above 10 atm while at 1 atm this value is about 30%. High partial pressure of CO_2 and H_2S or application of physical solvent with high absorption power leads to lower amounts of circulation. For example, the circulation rate need to absorption of CO_2 at a feed pressure of 10 atm in methanol is only about one-fourth of that circulation rate under 1 atm operating pressure.

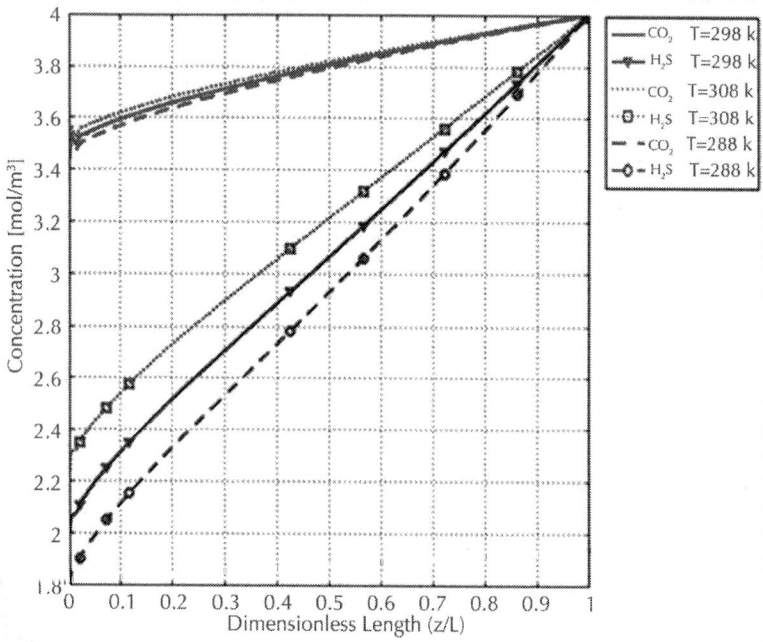

Figure 11: Effect of temperature on CO_2 and H_2S shell side concentration profile for V_L = 1 m/s, V_G = 3 m/s, C_{0,CO_2} = C_{0,H_2S} = 4 mol/m³ and T = 298 K.

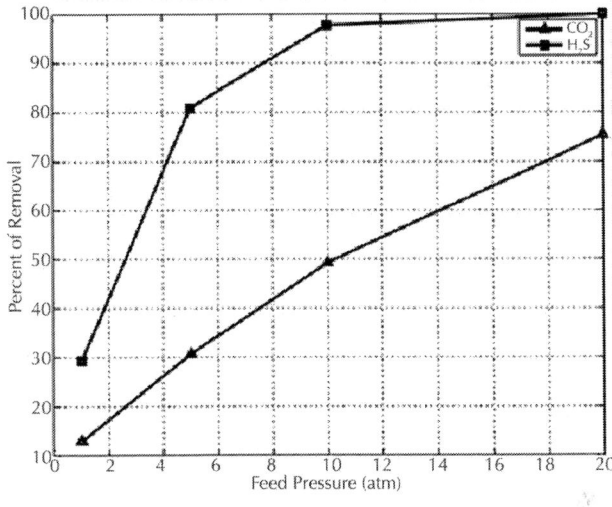

Figure 12: Effect of pressure on CO_2 and H_2S removal efficiencies for Methanol absorbent at $V_L = 0.1$ m/s, $V_G = 3$ m/s and T = 298 K.

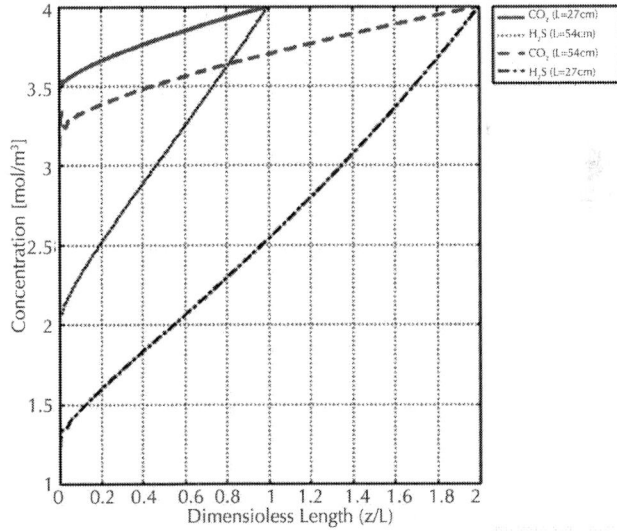

Figure 13: Effect of fiber length on CO_2 and H_2S concentration depletion in the gas phase along fiber length for methanol absorbent at $V_L = 1$ m/s, $V_G = 3$ m/s and T = 298 K.

Note that because the model results are based on "nonwetted mode" in which the gas mixture filled the membrane pores and assuming complete non-wetting conditions is valid at low pressure operations, it is expected that model overestimates CO_2 removal efficiency at high pressure. This can be attributed to mass transfer resistance caused by partial wetting of pores in hollow fiber membrane [5].

The membrane length needed to achieve the desired removal efficiency is a significant parameter. By increasing the membrane length, the membrane area for mass transfer increased and thus, higher removal efficiency is achieved. As shown in Figure 13, we examined this effect for two membrane length, i.e. 27 cm and 54 cm. As a result of doubling the length, CO_2 removal efficiency increased by 60% and H_2S removal efficiency by 40% for V_L = 1 m/s, V_G = 3 m/s, C_{0,CO_2} = C_{0,H_2S} = 4 mol/m³ and T = 298 K.

CONCLUSIONS

The physical absorption of CO_2 and H_2S from $CO_2/H_2S/CH_4$ mixture (when the partial pressure of components is 10% of total pressure) was simulated. The effect of liquid velocity, gas velocity, temperature and pressure on removal efficiency was explored and the concentration distributions inside the shell, through the membrane, and within the tube side were studied. The results indicate that methanol has the potential as a low-cost, green physical solvent for CO_2 capture in HFMGA. Relative absorption rate of CO_2 using methanol absorbent is in the range 1.5 to 2.7 and relative absorption rate of H_2S using methanol absorbent is in the range 2.9 to 3.75 in comparison with the case of water absorbent for given operating conditions (V_G = 2 m/s, C_{0,CO_2} = C_{0,H_2S} = 4 mol/m³, T = 298 K). However, in simultaneously absorption of CO_2 and H_2S using methanol in MGA selectivity remains relatively constant with increasing the methanol absorbent flow rate. With increasing the temperature, the removal efficiencies slightly decreased. At high pressures, methanol as a physical solvent is more efficient and it might be an alternative to chemical solvents. Moreover, CO_2 removal efficiency about 60% and H_2S removal efficiency about 40% increased as a result of doubling the membrane length.

COMMEMORATION

The authors are particularly grateful to Prof. Mohammad Khoshnoodi from University of Sistan and Baluchestan, the project manager who passed away unexpectedly, for his suggestions and knowledge shared.

REFERENCES

1. R. Faiz and M. Al-Marzouqi, "Mathematical Modeling for the Simultaneous Absorption of CO_2 and H_2S Using MEA in Hollow Fiber Membrane Contactors," Journal of Membrane Science, Vol. 342, No. 1-2, 2009, pp. 269-278. doi:10.1016/j.memsci.2009.06.050
2. R. N. Maddox, "Gas Conditioning and Processing," Campbell Petroleum Series, 3rd Edition, Vol. 4, 1982.
3. L. Sumin, et al., "The Enhancement of CO_2 Chemical Absorption by K_2CO_3 Aqueous Solution in the Presence of Activated Carbon Particles," Chinese Journal of Chemical Engineering, Vol. 15, No. 6, 2007, pp. 842-846. doi:10.1016/S1004-9541(08)60012-9
4. A. Mandowara and P. K. Bhattacharya, "Membrane Contactor as Degasser Operated under Vacuum for Ammonia Removal from Water: A Numerical Simulation of Mass Transfer under Laminar Flow Conditions," Computers and Chemical Engineering, Vol. 33, No. 6, 2009. pp. 1123-1131. doi:10.1016/j.compchemeng.2008.12.005
5. A. Gabelman and S. T. Hwang, "Hollow Fiber Membrane Contactors," Journal of Membrane Science, Vol. 159, No. 1-2, 1999, pp. 61-106. doi:10.1016/S0376-7388(99)00040-X
6. V. Y. Dindore, D. W. F. Brilman and G. F. Versteeg, "Modelling of Cross-Flow Membrane Contactors: Mass Transfer with Chemical Reactions," Journal of Membrane Science, Vol. 225, No. 1-2, 2005, pp. 275-289. doi:10.1016/j.memsci.2005.01.042
7. H. Jeon, et al., "Absorption of Sulfur Dioxide by Porous Hydrophobic Membrane Contactor," Desalination, Vol. 234, No. 1-3, 2008, pp. 252-260. doi:10.1016/j.desal.2007.09.092
8. K. A. Hoff, et al., "Modeling and Experimental Study of Carbon Dioxide Absorption in Aqueous Alkanolamine Solutions Using

a Membrane Contactor," Industrial & Engineering Chemistry Research, Vol. 43, No. 16, 2004, pp. 4908-4921. doi:10.1021/ie034325a

9. A. Kohl and R. Nielsen, "Gas Purification," 5th Edition, Gulf Publishing Company, Houston, 1997.

10. G. Hochgesand, "Rectisol and Purisol," European and Japanese Chemical Industrials Symposium, 1970, Vol. 62, No. 7, pp. 37-43.

11. J. A. Delgado, et al., "Simulation of CO_2 Absorption into Aqueous DEA Using a Hollow Fiber Membrane Contactor: Evaluation of Contactor Performance," Chemical Engineering Journal, Vol. 62, 2009, pp. 396-405. doi:10.1016/j.cej.2009.04.064

12. R. Wang, D. F. Li and D. T. Liang, "Modeling of CO_2 Capture by Three Typical Amine Solutions in Hollow Fiber Membrane Contactors," Chemical Engineering and Processing, Vol. 43, No. 7, 2004, pp. 849-856. doi:10.1016/S0255-2701(03)00105-3

13. W. Rongwong, R. Jiraratananon and S. Atchariyawut, "Experimental Study on Membrane Wetting In Gas-Liquid Membrane Contacting Process for CO_2 Absorption by Single and Mixed Absorbents," Separation and Purification Technology, Vol. 69, 2009, pp. 118-125.doi:10.1016/j.seppur.2009.07.009

14. D. Wang, W. K. Teo and K. Li, "Removal of H_2S to Ultra-Low Concentrations Using an Asymmetric Hollow Fibre Membrane Module," Separation and Purification Technology, Vol. 27, No. 1, 2002, pp. 33-40. doi:10.1016/S1383-5866(01)00186-1

15. P. Keshavarz, J. Fathhikalajahi and S. Ayatollahi, "Mathematical Modeling of the Simultaneous Absorption of Carbon Dioxide and Hydrogen Sulfide in a Hollow Fiber Membrane Contactor," Separation and Purification Technology, Vol. 63, No. 1, 2008, pp. 145-155. doi:10.1016/j.seppur.2008.04.008

16. S. Wang, K. Hawboldt and M. A. Abdi, "Novel DualMembrane Gas-Liquid Contactors: Modelling and Concept Analysis," Industrial & Engineering Chemistry Research, Vol. 45, No. 23, 2006, pp. 7882-7891. doi:10.1021/ie051368d

17. A. Mansourizadeh, A. F. Ismail and T. Matsuura, "Effect of Operating Conditions on the Physical and Chemical CO_2 Absorption through the PVDF Hollow Fiber Membrane Contactor," Journal

of Membrane Science, Vol. 353, No. 1-2, 2010, pp. 192-200. doi:10.1016/j.memsci.2010.02.054

18. R. Faiz and M. Al-Marzouqi, "CO_2 Removal from Natural Gas at High Pressure Using Membrane Contactors: Model Validation and Membrane Parametric Studies," Journal of Membrane Science, Vol. 365, No. 1-2, 2010, pp. 232-241.doi:10.1016/j.memsci.2010.09.004

19. J. Happel, "Viscous Flowrelative to Arrays of Cylinders," AIChE Journal, Vol. 5, No. 2, 1959, pp. 174-177. doi:10.1002/aic.690050211

20. M. Al-Marzouqi, et al., "Modeling of CO_2 Absorption in Membrane Contactors," Separation and Purification Technology, Vol. 59, No. 1, 2008, pp. 286-293.

21. M. Mavroudi, S. P. Kaldis and G. P. Sakellaropoulos, "Reduction of CO_2 Emissions by a Membrane Contacting Process," Fuel, Vol. 82, No. 15-17, 2003, pp. 2153-2159.doi:10.1016/S0016-2361(03)00154-6

22. P. Keshavarz, J. Fathhikalajahi and S. Ayatollahi, "Analysis of CO_2 Separation and Simulation of a Partially Wet-Ted Hollow Fiber Membrane Contactor," Journal of Hazardous Materials, Vol. 152, No. 3, 2008, pp. 1237-1247. doi:10.1016/j.jhazmat.2007.07.115

23. G. F. Versteeg and W. P. M. Van Swaaij, "On the Kinetics between CO_2 and Alkanolamines both in Aqueous and Non-Aqueous Solutions I. Primary Andsecondary Amines," Chemical Engineering Science, Vol. 43, No. 3, 1988, pp. 573-585. doi:10.1016/0009-2509(88)87017-9

24. J. J. Carroll and A. E. Mather, "The Solubility of HydroGen Sulphide in Water from 0°C to 90°C and Pressure to 1 MPa," Geochimica et Cosmochimica Acta, Vol. 53, No. 6, 1989, pp. 1163-1170. doi:10.1016/0016-7037(89)90053-7

25. K. Lunsford and G. Mcintyre, "Decreasing Contactor Temperature Could Increase Performance," GPA Annual Convention, Bryan Research and Engineering, Inc., Texas, 1999, pp. 121-127.

26. E. L. Cussler, "Diffusion Mass Transfer in Fluid Systems," Cambridge University, Cambridge, 1984.

27. M. V. Diaz and A. Coca J., "Correlation for the Estimation of Gas-Liquid Diffusivity," Chemical Engineering Communications, Vol. 52, No. 4-6, 1987, pp. 271-281.doi:10.1080/00986448708911872

28. R. H. Perry, "Perry's Chemical Engineers' Handbook," 7th Edition, McGraw-Hill, New York, 1997.
29. B. E. Poling, J. M. Prausnitz and J. P. O'Connell, "The Properties of Gases and Liquids," 5th Edition, McGrawHill, New York, 2004.
30. R. Faiz and M. Al-Marzouqi, "H_2S Absorption via CarBonate Solution in Membrane Contactors: Effect of Species," Journal of Membrane Science, Vol. 350, No. 1-2, 2010, pp. 200-210. doi:10.1016/j.memsci.2009.12.028
31. V. Y. Dindore, D. W. F. Brilman and G. F. Versteeg, "Hollow Fiber Membrane Contactor as a Gas-Liquid Model Contactor," Chemical Engineering Science, Vol. 60, No. 2, 2005, pp. 467-479. doi:10.1016/j.ces.2004.07.129

Citations

CHAPTER 1

Mohammad Ali Khodagholi and Mohammad Irani, "Catalytic and Noncatalytic Conversion of Methane to Olefins and Synthesis Gas in an AC Parallel Plate Discharge Reactor," Journal of Chemistry, vol. 2013, Article ID 676901, 7 pages, 2013. doi:10.1155/2013/676901.

CHAPTER 2

Yuanchen Zhu, Travis Robinson, Amani Al-Othman, André Y. Tremblay, and Marten Ternan, "n-Hexadecane Fuel for a Phosphoric Acid Direct Hydrocarbon Fuel Cell," Journal of Fuels, vol. 2015, Article ID 748679, 9 pages, 2015. doi:10.1155/2015/748679.

CHAPTER 3

Kongzhai Li, Hua Wang, and Yonggang Wei, "Syngas Generation from Methane Using a Chemical-Looping Concept: A Review of Oxygen Carriers," Journal of Chemistry, vol. 2013, Article ID 294817, 8 pages, 2013. doi:10.1155/2013/294817.

CHAPTER 4

Júnior, L., Silva, A., Silva, B. and Alencar, S. (2014) Synthesis of ZSM-22 in Static and Dynamic System Using Seeds. Modern Research in Catalysis, 3, 49-56. doi: 10.4236/mrc.2014.32007.

CHAPTER 5

Shadi Vafaeyan, Alain St-Amant, and Marten Ternan, "Nickel Alloy Catalysts for the Anode of a High Temperature PEM Direct Propane Fuel Cell," Journal of Chemistry, vol. 2014, Article ID 151638, 8 pages, 2014. doi:10.1155/2014/151638.

CHAPTER 6

Sivakumar VM, Abdul Rahman Mohamed, Ahmad Zuhairi Abdullah, and Siang-Piao Chai, "Role of Reaction and Factors of Carbon Nanotubes Growth in Chemical Vapour Decomposition Process Using Methane—A Highlight," Journal of Nanomaterials, vol. 2010, Article ID 395191, 11 pages, 2010. doi:10.1155/2010/395191.

CHAPTER 7

Christos M. Kalamaras and Angelos M. Efstathiou, "Hydrogen Production Technologies: Current State and Future Developments," Conference Papers in Energy, vol. 2013, Article ID 690627, 9 pages, 2013. doi:10.1155/2013/690627.

CHAPTER 8

Edem Cudjoe Bensah and Moses Mensah, "Chemical Pretreatment Methods for the Production of Cellulosic Ethanol: Technologies and Innovations," International Journal of Chemical Engineering, vol. 2013, Article ID 719607, 21 pages, 2013. doi:10.1155/2013/719607.

CHAPTER 9

D. Kwak, M. Kim, J. Kim, Y. Oh, S. Noh, B. So, S. Jung, S. Jung and S. Chae, "Evaluation of Methane Yield on Mesophilic-Dry Anaerobic Digestion of Piggery Manure Mixed with Chaff for Agricultural Area," Advances in Chemical Engineering and Science, Vol. 3 No. 4, 2013, pp. 227-235. doi: 10.4236/aces.2013.34029.

CHAPTER 10

M. Mahdavian, H. Atashi, M. Zivdar and M. Mousavi, "Simulation of CO2 and H2S Removal Using Methanol in Hollow Fiber Membrane Gas Absorber (HFMGA)," Advances in Chemical Engineering and Science, Vol. 2 No. 1, 2012, pp. 50-61. doi: 10.4236/aces.2012.21007.

Index

A

Activated carbons (ACs) 116
Aqueous phase reforming (APR) 152
Autothermal reforming (ATR) 141, 144

C

Carbon blacks (CBs) 116
Carbon Nanotubes (CNT) 108
Catalyst layers (CL) 27
Catalytic 3, 13, 18
Catalytic chemical vapour decomposition (CCVD) 109
Chemical looping combustion\" (CLC) 48
Chemical-looping selective oxidation of methane (CLSOM) 46, 58
Chemical vapour deposition (CVD) 108
Computational mass transfer (CMT) 248

D

Density functional theory 87, 89, 104, 105
Dimethyl ethers of polyethylene glycol (DEPG) 250
Direct hydrocarbon fuel cells (DHFCs) 22

E

Exponential Model (EM) 235

F

Fossil fuel 141

G

Gas diffusion layers (GDL) 27
Generalized gradient approximation (GGA) 89

H

Hollow fiber membrane gas absorber (HFMGA) 247
Hot filament (HF) 112
Hydrocarbons 2, 3, 4, 10, 13, 14, 16, 17, 18
Hydroxymethyl-furfural (HMF) 167

I

Infrared region (IR) 73

L

Lattice constant (LC) 90
Lignocellulosic biomass (LB) 166
Low-pressure thermal chemical deposition (LPTCD) 113

M

Membrane electrode assembly (MEA) 22, 27
Membrane fuel 86, 104
Modified Gompertz Model (MGM) 235

O

Outer diameter 1, 4

P

Partial oxidation of methane (POM) 46
Perdew-Burke-Ernzerhof (PBE) 90
Phosphoric acid fuel cell (PAFC) 22, 26
Proton exchange membrane (PEM) 146, 154
Pulsed laser vaporization technique (PLV) 109

S

Simultaneous saccharification and fermentation (SSF) 172, 206
Single-walled carbon nanotubes (SWNTs) 108
Solid oxide electrolysis cells (SOEC) 154
Steam reforming (SR) 141
Stoichiometric ratio (SR) 27
Syngas generation 46

V

Volatile solid (VS) 224, 241

W

Water-gas shift reaction (WGS) 152
Water-gas shift (WGS) 141, 145

X

X-ray diffraction (XRD) 72, 73